Hydrogen

Hot Stuff
Cool Science

Hydrogen
Hot Stuff
Cool Science

Journey to a World of
Hydrogen Energy and Fuel Cells
at the Wasserstoff Farm

Rex A. Ewing

HYDROGEN—Hot Stuff Cool Science:
Journey to a World of Hydrogen Energy and Fuel Cells at the Wasserstoff Farm

Published by PixyJack Press, LLC
PO Box 149, Masonville, CO 80541 USA
www.PixyJackPress.com

First Edition 2004

9 8 7 6 5 4 3 2 1

ISBN 0-9658098-6-2

Library of Congress Cataloging-in-Publication Data
Ewing, Rex A.
Hydrogen--hot stuff cool science : journey to a world of hydrogen energy and fuel cells at the Wasserstoff Farm / Rex A. Ewing.-- 1st ed.
p. cm.
Includes bibliographical references and index.
ISBN 0-9658098-6-2
1. Hydrogen as fuel. 2. Hydrogen as fuel--Research. I. Title.

TP359.H8E95 2004
665.8'1--dc22

2004016708

Printed in the U.S.A. on ECF (Elemental Chlorine Free) paper with soy ink.

Distributed to the trade by Johnson Books, Boulder, Colorado.
1-800-258-5830 www.johnsonbooks.com

Book design and technical illustrations by LaVonne Ewing.
All other illustrations by Sara Tuttle.

For LaVonne Ann
you make it all seem so easy

table of contents

continued

preface

★ ★ ★

When I was asked to write a book about hydrogen energy, I naively assumed it would be an easy task. It was not because I thought hydrogen energy would be an unchallenging subject. To the contrary, from the very beginning I had at least *some* appreciation of its scope and complexities. No, I thought it would be easy for the simple reason that I imagine *anything* I haven't yet done will be easy. This is how I talk myself into things; it's a personality trait I share with Daffy Duck. Even the doom-and-gloom, worst-case-scenario trolls lurking deep within my psyche failed to burrow their way into consciousness as I was deliberating whether or not to take on the project. I should have known then and there, by the trolls' conspiratorial lethargy, that I might just be kidding myself. But somehow I didn't. Perhaps it was fate.

However it came about, I now confess that this is the hardest book I've ever written. Its difficulty arises not from the complexity of the subject's individual parts, but from the expansive nature of the whole animal. Writing about hydrogen energy is, in many ways, writing about the future history of the world in all of its intricacies, since most of the technologies explored—and I use that word in a literal sense—are designed to forever transform the way we interact with the planet we live on.

But it's also been the most fun book I've ever written, because *Hydrogen—Hot Stuff Cool Science* is, more than anything else, an adventure. It's an adventure into the science of hydrogen energy, certainly, but it's also an adventure in its own right. This was nothing I planned, really—more just something that happened along the way from the first page to the last, as I was taken under the wing of an eccentric wizard in a fanciful place called the Wasserstoff Farm. (I'd like to say more here, but I'm already in danger of exposing the plot.)

For readers with solid science backgrounds, the first few chapters may seem deceptively simple. They are included mainly for those of you whose most vivid memory of Chemistry class is the pustule-like mole on Professor Schlitchenbarker's forehead. Or for those who may need a little brushing up on the not-unimportant distinction between breaking atomic bonds, and breaking atoms. But you'll also learn how and why the atmosphere resembles a greenhouse, why we're fortunate it does, and how to avoid getting too much of a good thing. And to make sure we all know just how big of a problem we're facing, a visually thought-provoking analysis of how much energy the U.S. actually consumes every year polishes off the first section.

Then it's on to the meat of the story: hydrogen energy. Where hydrogen comes from, the myriad ways to liberate it, and what to do with it, once we have it. The proposals for making hydrogen are as pristine and spritely as sunlight and water, as airy as a Midwest breeze, as down-and-dirty as coal, or as darkly forbidding as radioactive decay.

Making hydrogen, of course, is just a third of the equation, for we then have to decide how best to store something so light that a 5-gallon bucket of the stuff—even chilled and compressed into liquid form—weighs no more than a goose-down pillow. Do we squeeze it into ever smaller volumes (hydrogen becomes downright recalcitrant above 10,000 psi), or hide it inside microscopic tubes of carbon? Or is 200-proof moonshine the ideal way to get hydrogen from here to there?

And finally, what's the best way to use hydrogen? Should we hold out for the promise of fuel cells, or burn it in modified V-8s?

Since it's not always possible to be at once colorful and attentive to detail, you will find a reference section of a technical nature following every chapter. More than a simple listing of sources, or endnotes (a referencing scheme for which I have always had a strong aversion), the *Technistoff* sections provide specific data and resources to buttress the chapters' broad-based lessons.

The concept of hydrogen energy is so intrinsically beautiful it would seem a crime to burden it with the yoke of political rhetoric. I will, therefore, leave that muddy task to those who wear higher boots than I do. I learned long ago my political talents were as stunted as my piano-playing abilities, so I refrain from practicing either. Watching old Arnold Schwarzenegger movies is about as close as I ever get to hydrogen politics.

And finally: aside from the excellent graph of projected costs of hydrogen production on page 265, you will not find endless tables and graphs projecting capital costs, returns on investment, or the possible future cost of *x* if, and only if, conditions *a*, *b*, *c* and *d* are simultaneously satisfied. I'm just a humble purveyor of scientific knowledge, not an economist. But I do know that if we want something bad enough, we'll find a way to pay for it. Always have, always will.

Besides, as I may have already said, this book is an *adventure*.

REX A. EWING
August 2004

prologue

★ ★ ★

THE WASSERSTOFF FARM

They say necessity is the mother of invention. Whenever I consider the climate of circumstances surrounding the few inventions I've been responsible for, I'd have to agree that they (whoever "they" may be) are right; I really *did* need most of the peculiar gadgets and crude task-specific tools that survived the *ad hoc* process from vague mental machinations to physical fabrication, and slid off my workbench, useful and functional. But that's not even half the story. Necessity, I've learned, has a cousin. Her name is Discovery, and Necessity's sister, Serendipity, is Discovery's nursemaid.

Serendipity has been whimsically defined as looking for a needle in a haystack, and finding instead the farmer's daughter. It's that, and more. It's throwing a pickaxe, in utter frustration, at the wall of a woefully unproductive mine and dislodging the rock that hid the Mother Lode. Or stumbling through the woods in search of food, tripping on a log, sliding down a hill, and landing face-first in a berry patch.

It's luck so good it can't be dismissed as pure chance.

And so it was the day I happened upon the Wasserstoff Farm.

I had been planning to write a book about hydrogen. Usually when I'm preparing to write a book, I spend months stuffing my head with information. I stuff and cram and shake and push and tamp until my head gets so full of facts my fingers begin to itch. Then I feverishly excavate my desk until a keyboard appears beneath a mountain of books, periodicals and research papers, whereupon I sit down to write. At least it usually works that way.

This time was different. Although my head was ready to burst, I still didn't feel that old familiar itch in my fingers. After a little reflection, the nature of the problem became clear—my head was indeed full, as it should have been at that stage of the game, but it was more full of questions than knowledge. I felt as though I knew less than I did back before I didn't know anything at all.

The study of hydrogen—and by extension, chemistry—has a way of doing that to people.

So I gave up, for the day at least. After all, it was a warm and cloudless day in late fall—the kind of day you cherish simply because you know there are so few of them left on the year's agenda. It was a day that begged for a ride in the country. I first thought of taking my pickup up the canyon to the top of the pass, but, after casting my eyes toward the ugly brown cloud of exhaust fumes and industrial pollution that hovered over the distant city below me, I decided it might make more sense to go on a horseback ride through the backwoods.

My better judgment told me to take Blackie, a tried-and-true 10-year-old gelding; my sense of adventure prodded me to saddle up Mike, instead. Mike was a half-broke 2-year-old colt with the raw energy of a tornado mated to the cerebral wherewithal of a jack rabbit with its tail on fire. With Mike it would be anything but a lazy, peaceful ride through the enchanted forest. On the other hand, it was a nice day, Mike needed the work, and the ground—where, by my calculations, I had a better than average chance of ending up—still wasn't frozen.

So off we went, me and Mike.

The first couple of miles were uneventful. We threaded our way along a steep-walled creek bed to where it opened up into a high grassy meadow. I paused to take in the scene before me. Mike had been on his best behavior to that point, neither bolting, jumping, or crow hopping, but I couldn't get over the feeling I was straddling a keg of dynamite. I talked softly to him, gently stroking his strong, tense neck. It helped a little, but I could still feel his muscles twitch under the saddle as he chewed nervously at the bit. His quick, darting eyes were the size of silver dollars and every little sound—a bird flying out of a bush; the rustling of the dry grass in the gentle breeze—elicited a sharp turn of his head. I feared it was just a matter of time before we had a rodeo on our hands. Like hydrogen and oxygen trapped in the same bottle, all that was required was a spark to bring our peaceful coexistence to an abrupt and stormy end.

My only hope was that the catalyst that finally set him off was a moderate one; I was hoping to control the reaction, so to speak. A wind-blown weed would have been manageable, or even a herd of deer breaking into the clearing in the distance. Something told me I wasn't going to be so lucky.

I pressed my heels into his sides and clucked a couple of times, urging him on, hoping to keep his mind off whatever spooks and goblins lay hidden in the grass and brush around us. Mike relaxed a bit, happy to be moving again, and by the time we reached the end of the clearing I was half convinced we were going to make it through the ride without incident.

In retrospect, it was a naïve conclusion.

A terrible racket erupted as we began to ease back into the forest. Mike jumped quickly back and spun to the side as a flock of wild turkeys flew out of the trees, a hungry coyote hot on their tails. Neither the turkeys nor the coyote were in the least deterred at the sight of us, but the spirited colt was not nearly so bold. Here, at last, were the demons he'd known all along were lurking in the shadows and he wasn't going to waste a second in getting away from them. There was but a single narrow trail leading into the darkness of the heavily wooded forest before us and Mike shot toward it like a bolt from a crossbow.

What exactly happened after that is a little unclear. I remember being slapped and swatted by low hanging branches while trying to bring Mike about. I wasn't having much luck; a little less than none, actually. The colt was not slowed in the least by the thick, horizontal foliage that all seemed to be just a little higher than his head and a little lower than mine.

I ducked and bobbed to no avail. Every time I sat up to yank on the reins, I'd have to flatten out again to keep from getting brained. It was a nightmare. Finally, after getting clobbered so hard by a wickedly-placed aspen branch I thought I was going to pass out, I lay down in the saddle and waited for Mike to run himself out.

After what seemed like an hour—though it certainly could not have been more than a few minutes—he finally stopped; a bit too abruptly. The horse planted his feet in the ground and leaned back while I slid right over his head, did a half-flip in the air, and landed squarely on my backside in front of him.

My head throbbed, my tailbone ached, and little arrows of pain stabbed at me everywhere in between. But, to my relief, everything seemed to work and I was off the horse, so I had at least achieved my objective, however ungracefully.

But where was I? As my head began to clear and my eyes came back into focus, it became obvious that my supercharged colt had taken me to a place I'd never been before, for there in front of me was the biggest stone wall I'd ever seen. Looking to either side I could see no end to it. Judging by the nearby trees, it was at least 70 feet from the ground to the crenels

lining the top. Vines clung to it from top to bottom. Lichens had largely covered the perfectly-placed, irregularly-shaped stones, and little sprigs of foliage grew here and there from the crumbling joints between them. It looked as though it had been there for a thousand years.

A heavy wooden double gate with a rounded top and stout iron hinges provided the only break anywhere along the massive wall. Over the top of the gate three words had been expertly carved into the header stone in bas relief:

The Wasserstoff Farm

I must've taken a harder lick on the head than I thought.

Certain that I was hallucinating, and therefore eager to put as much distance as possible between myself and my trauma-induced illusions, I gathered myself together and managed to get my feet underneath me. Mike was a few paces away, peacefully munching on the tall grass growing in front of the wall. I grabbed the reins, hopped on his back, and said, "Let's go, hot rod, and no tricks this time."

I had just turned the horse about when we were both startled by an unpleasantly discordant sound. Mike spun around and we both watched wide-eyed as one of the giant doors slowly opened on its creaky hinges. The fact that Mike was reacting to the scene unfolding before us was a good indication that I really wasn't hallucinating, unless, of course, I was also hallucinating Mike's reaction, in which case I was in even more trouble than I thought. As I mulled this over, a man appeared from beyond the partially opened gate. He looked like the younger brother of Father Time. "What's your hurry?" he asked, with a merry smile. "I've been expecting you. Come in, come in! Welcome to the Wasserstoff Farm!"

I considered turning and heading for home, but something compelled me, instead, to touch my heels against Mike's flank and follow the old gent inside.

book one

★

LAYING THE FOUNDATION

Atoms H2O
Protons Neutrons
Nucleus Electrons Orbits
Bonds Molecules Isotopes
Organic Compounds Carbon
Oxygen Hydrogen Nitrogen
Helium Methanol Ethanol
Carbon Dioxide Methane
Atmosphere Alcohols
Global Warming O2
Hydrocarbons
Ethane

chapter 1

Meet the Wizard at the Wasserstoff Farm

As if by magic, the gate closed behind us as we followed the odd fellow inside the walls of the Wasserstoff Farm. Instinctively, I took stock of my strange, new environment. Mike did the same, though with far more trepidation, it seemed, than curiosity.

Once inside the gate, the giant wall bore scant resemblance to the outside; devoid of vines and lichens, the enormous stones were as shiny and smooth as top-dollar headstones and the joints between them were surgically precise. Unlike the forest I had just passed through, aspen and spruce trees here grew in distinct groves, or copses, like flowers in a garden. Rock outcroppings were likewise in discrete loci. They looked natural enough, but I had the impression they had been placed in a meaningful pattern, like Stonehenge, though far subtler.

The wall of the compound—a word I use with some misgiving—seemed to extend, more or less on level ground, to beyond the horizon in every direction. I noticed that it had a slight curvature, as if I were inside a giant circular enclosure. I could see no end to the curved wall, nor could I see any walls adjoining it. This was, of course, impossible, since I knew for a fact that just outside the gate there was a heavily-wooded valley closed in by mountains on either side.

Even more interesting than the natural-yet-orderly landscape were the buildings. There was nothing consistent about them; some were stone, others wood or glass, or a combination of all three. Some were sizable edifices, others little more than sheds. They were sparsely placed, yet due to the vastness of the place, there were several within my field of view. All were connected by stone pathways.

It was all more than I could ever hope to fathom. I made a mental note to pat myself on the back for dreaming up such a grand illusion.

Later. Now it was time to get on with seeing who this man was and what this place was all about. I slid off Mike's back and tied the reins behind his neck. Suddenly deciding that the grass underfoot was more compelling than the strangeness all about him, he dropped his head to the ground and began pulling out lush, green tufts in succulent mouthfuls.

I turned to face the old man, who had been patiently waiting for me—and Mike—to familiarize ourselves with the new surroundings. He was a tall man, and slender, though hardly frail or insubstantial. To the contrary: he looked as quick and wiry as a mountain cat. His long, flowing white hair gave the impression of age, but his lean face was timeless. It seemed as soft as baby flesh and without whiskers. The wrinkles that appeared and vanished with his changing facial expressions seemed more to enhance his character, than to offer any testament of life's hardships. Nor did his deep-blue eyes betray his age. They conveyed the wisdom that one associates with advanced years, yet at the same time they sparkled with the mischievousness of youth. I could see he was studying me, as I studied him.

Finally, he spoke. "My name is Zedediah. Zedediah Pickett, if you're one for last names. And this is, as I've said, the Wasserstoff Farm."

Pickett. The name was vaguely familiar, but I couldn't place it. And wasn't *Wasserstoff* the German word for Hydrogen?

"Rex," I said, extending my hand. "Rex Ewing."

"Of course you are." He took my hand as if the gesture was a distraction and shook it. A firm handshake, despite his reluctance.

He didn't waste any time getting to the heart of the matter. He said, "Rumor has it you're going to write a book about Wassersto...I mean, hydrogen." He rested his chin in the fork between his index finger and thumb and regarded me with an impish stare. "It's rather an expansive subject, don't you know?"

I didn't bother to ask how he'd heard such a rumor, since it didn't appear that he got out much. I simply answered, "Well, yes, I realize that." I had the distinct feeling I was getting myself into something I'd not soon be getting out of.

"Not much happens in life without it," he added.

"No, I s'pose not."

"Going to be a large book, I'd say. You just might be an old man by the time you finish it." With his finger he traced out a giant rectangle in the air. To my amazement, the area inside the rectangle turned a ghostly, translucent black with thick, white edges. It soon resolved into a huge, ethereal book. I was trying to read the title when, with a wave of his hand, the book disappeared. "You're going to need some help," he said, somewhat sympathetically.

"Actually, it's just going to be about hydrogen as fuel," I answered lamely.

He laughed. "Just? *Just?* Listen, lad, you could spend the next fifty years describing all the ways hydrogen is used as fuel. Why, the reactions that take place inside the body of a gnat are enough to turn your hair white." He pulled on his long, white locks and laughed again. As he did, a gnat appeared out of nowhere and quickly landed on the back of my hand. I swatted it instinctively, but when I removed my hand there was no trace of it.

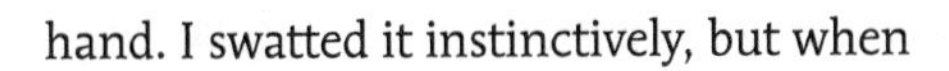

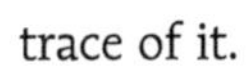

"What I meant to say," I answered, feeling a little frustrated, "as fuel for cars, and other things. You know, fuel cells and all that."

"Oh...well, that's a bit more manageable. But what about all the problems in getting the stuff in the first place? And then storing it? And don't forget transportation and distribution. It's all tied together, I'm afraid."

"Sure, but–"

He cut me off, stabbing the sky with a long, gracile finger. "And then there's carbon! And oxygen! Can't talk about hydrogen without carbon and oxygen. Isn't carbon dioxide the bad boy of the planet, causing all this clamor about the 'hydrogen economy'? Good Heavens! To hear people talk you'd think the Earth is covered in a crust of the carbon three feet thick. You'd think there was so much of it in the atmosphere that it was crowding out all the oxygen and making it hard to breathe."

"Well, there *is* a lot of it–"

"Poppycock! There's hardly any. There's over 560 times more oxygen in the atmosphere than carbon dioxide, and 2,100 times more nitrogen. Why even argon, that stoic, standoffish gas that refuses to associate with the rest of nature, is 25 times more plentiful than carbon dioxide. And down here on terra firma it's not much different. Did you know there's over 25 times more titanium–titanium of all things!–in the Earth's crust than there is carbon? And over 400 times more aluminum?"

"Well then," I said, "if there's so little of it, it must not be the problem everyone thinks it is."

Zedediah shook his head and looked at me with a worrisome combination of resignation and pity. Slowly he answered, "Now think this through a minute, lad. If you have a whole bunch of something, you have to add a lot to it to make it seem like you've got more, right? Say, for instance, that you have 1,000 apples. If someone hands you two more, it doesn't much matter. You'd have to count them all to even know the dif-

Zed sez

Did you know there's over 25 times more titanium—titanium of all things!—in the Earth's crust than there is carbon? And over 400 times more aluminum?

ference. But if you only have ten apples to begin with, then those extra two begin to look significant. By the same token, if the atmosphere was naturally 3 percent CO_2 then we could dig up all the carbon-based fuel we could find and burn it with impunity until the end of time, without ever having to give a hoot about this 'global warming' thing. But the atmosphere isn't 3 percent CO_2, it's only a paltry 0.037 percent..." He paused and looked at me quizzically, then asked, "Is any of this getting through to you?"

I'd never felt so stupid in my life. It was even worse than the time I'd tried to coax a tick out of my dog's underside with alcohol and match. I mumbled, "Uh...yes'r."

He smiled in a fatherly way, put his arm around my shoulders, and said, "Good! I'm sure you'll catch on in no time. And by the way, drop that 'sir' business. The name's Zed, just like that abrasive fellow in those *Men in Black* movies."

"You watch those silly things?" I asked in disbelief.

"Every chance I get, lad. A man can never have too much education."

For the first time since I'd set foot inside the Wasserstoff Farm, I was beginning to think Zedediah Pickett might actually be for real.

"Anyway," he continued, "by the looks of it, you've had a hard day. Better get some rest. We'll start your instruction first thing in the morning. Your room is right over there." He was pointing to a small stone cottage several yards away.

"Oh, I can't stay," I protested, "I've got things to tend to—"

"Trust me. Everything will be fine; just like you were never gone." With that, he turned and walked away.

I stared at the large, wooden gate I'd come through earlier. It was the only way out, that much was certain. But there was neither handle nor latch, and something told me it wasn't going to open with a push.

I shook my head and let out a sigh, then unsaddled my horse and retired to my cottage.

technistoff - 1

Thus began my encounter with Zedediah Pickett and the Wasserstoff Farm. Though I tried to contribute the whole episode to the many knocks to my head I'd taken on my wild ride, I was left with the task of explaining where all of Zed's numbers came from. These were not numbers I could have siphoned off the top (or any other part) of my head, because they were never there to begin with. I reasoned, therefore, that if Zed's figures were real, then Zed must also be real.

atmospheric gases

The abundance of atmospheric gases was easily found in several places in David L. Heiserman's, *Exploring Chemical Elements and their Compounds*, though Zed's assertion that atmospheric CO_2 concentrations are currently at 0.037 percent is more recent—and therefore more correct—than Heiserman's slightly lower estimate of 0.03 percent.

Concentrations of atmospheric gases, I later discovered, are always given as volume percentages (such as 0.037 for CO_2) or ppmv (parts per million by volume, which for CO_2 would be 370). That's because, as demonstrated by Amadeo Avogadro in 1811, equal volumes of different gases under the same conditions of pressure and temperature contain the same number of molecules. (This discovery assured that "Avogadro" would forever be a household word—at least in scientists' households—since the number of gas molecules in a mole, 6.022 x 10^{23}, is known universally as Avogadro's Number.)

earth's elements

Such is not the case with elements in the Earth's crust. Different elements take up vastly different volumes, as do different forms of the same element, as witnessed by the difference in density between coal and diamond. Therefore, when measuring percentages, or parts per million (ppm) of solid materials, mass (weight) is used, instead of volume.

It's easy to find references that list the ppm of the Earth's crust for the first ten most common elements (O, Si, Al, Fe, Ca, Na, Mg, K, Ti and H), but progressively harder after that. Fortunately, the website *http://www.scescape.net/~woods/* gives a wealth of information about every natural element, including the abundance of each element in the Earth's crust. You should be warned, however, that the number is given as a logarithm. To convert it to ppm you have to extract the log by using it as an exponent of 10. So, if they give the abundance of, say, chlorine as log 2.1, you can calculate $10^{2.1}$ for your answer (about 126 ppm). Any scientific calculator can handle this operation in a single keystroke.

Happily, all of Zed's numbers checked out.

List of Elements in Atomic Number Order

#	Symbol	Name	Atomic Weight
1	H	Hydrogen	1.0079
2	He	Helium	4.0026
3	Li	Lithium	6.941
4	Be	Beryllium	9.0122
5	B	Boron	10.811
6	C	Carbon	12.011
7	N	Nitrogen	14.007
8	O	Oxygen	15.999
9	F	Fluorine	18.998
10	Ne	Neon	20.180
11	Na	Sodium	22.990
12	Mg	Magnesium	24.305
13	Al	Aluminium	26.982
14	Si	Silicon	28.086
15	P	Phosphorus	30.974
16	S	Sulfur	32.065
17	Cl	Chlorine	35.453
18	Ar	Argon	39.948
19	K	Potassium	39.098
20	Ca	Calcium	40.078
21	Sc	Scandium	44.956
22	Ti	Titanium	47.867
23	V	Vanadium	50.942
24	Cr	Chromium	51.996
25	Mn	Manganese	54.938
26	Fe	Iron	55.845
27	Co	Cobalt	58.933
28	Ni	Nickel	58.693
29	Cu	Copper	63.546
30	Zn	Zinc	65.409
31	Ga	Gallium	69.723
32	Ge	Germanium	72.64
33	As	Arsenic	74.922
34	Se	Selenium	78.96
35	Br	Bromine	79.904
36	Kr	Krypton	83.798
37	Rb	Rubidium	85.468
38	Sr	Strontium	87.62
39	Y	Yttrium	88.906
40	Zr	Zirconium	91.224
41	Nb	Niobium	92.906
42	Mo	Molybdenum	95.94
43	Tc	Technetium	[98]
44	Ru	Ruthenium	101.07
45	Rh	Rhodium	102.91
46	Pd	Palladium	106.42
47	Ag	Silver	107.87
48	Cd	Cadmium	112.41
49	In	Indium	114.82
50	Sn	Tin	118.71
51	Sb	Antimony	121.76
52	Te	Tellurium	127.60
53	I	Iodine	126.90
54	Xe	Xenon	131.29
55	Cs	Caesium	132.91
56	Ba	Barium	137.33
57 - 71		Lanthanum Series	
72	Hf	Hafnium	178.49
73	Ta	Tantalum	180.95
74	W	Tungsten	183.84
75	Re	Rhenium	186.21
76	Os	Osmium	190.23
77	Ir	Iridium	192.22
78	Pt	Platinum	195.08
79	Au	Gold	196.97
80	Hg	Mercury	200.59
81	Tl	Thallium	204.38
82	Pb	Lead	207.2
83	Bi	Bismuth	208.98
84	Po	Polonium	[209]
85	At	Astatine	[210]
86	Rn	Radon	[222]
87	Fr	Francium	[223]
88	Ra	Radium	[226]
89 - 103		Actinides Series	
104	Rf	Rutherfordium	[261]
105	Db	Dubnium	[262]
106	Sg	Seaborgium	[266]
107	Bh	Bohrium	[264]
108	Hs	Hassium	[277]
109	Mt	Meitnerium	[268]

Periodic Table of Chemical Elements

1A	2A	3B	4B	5B	6B	7B	8B			1B	2B	3A	4A	5A	6A	7A	8A
1 **H**																	2 **He**
3 **Li**	4 **Be**											5 **B**	6 **C**	7 **N**	8 **O**	9 **F**	10 **Ne**
11 **Na**	12 **Mg**											13 **Al**	14 **Si**	15 **P**	16 **S**	17 **Cl**	18 **Ar**
19 **K**	20 **Ca**	21 **Sc**	22 **Ti**	23 **V**	24 **Cr**	25 **Mn**	26 **Fe**	27 **Co**	28 **Ni**	29 **Cu**	30 **Zn**	31 **Ga**	32 **Ge**	33 **As**	34 **Se**	35 **Br**	36 **Kr**
37 **Rb**	38 **Sr**	39 **Y**	40 **Zr**	41 **Nb**	42 **Mo**	43 **Tc**	44 **Ru**	45 **Rh**	46 **Pd**	47 **Ag**	48 **Cd**	49 **In**	50 **Sn**	51 **Sb**	52 **Te**	53 **I**	54 **Xe**
55 **Cs**	56 **Ba**	57-71 **Rare Earths**	72 **Hf**	73 **Ta**	74 **W**	75 **Re**	76 **Os**	77 **Ir**	78 **Pt**	79 **Au**	80 **Hg**	81 **Tl**	82 **Pb**	83 **Bi**	84 **Po**	85 **At**	86 **Rn**
87 **Fr**	88 **Ra**	89-103 **Acti-nides**	104 **Rf**	105 **Db**	106 **Sg**	107 **Bh**	108 **Hs**	109 **Mt**									

Not Shown: Lanthanides 58 - 71 (Rare Earth elements) and Actinides 90 - 103

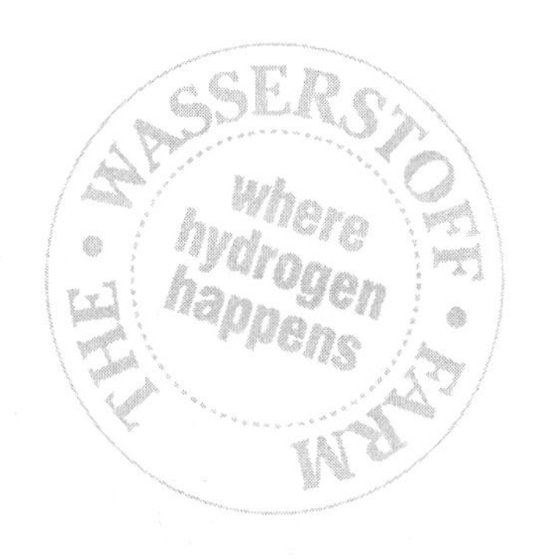

chapter 2

The Birth of Hydrogen: On A Fine Day, 13.7 Billion Years Ago...

Zed rousted me out of bed a little before sunrise. I was dreaming that I'd been kidnapped by scrawny, cow-eyed aliens and whisked away to the planet Wasserstoff, where I was forced to build a starlight-powered hydrogen electrolysis plant. When I opened my eyes and saw Zed, I realized my dream wasn't that far off the mark.

"Get up! Get up!" he implored. "The early worm avoids the bird...or however that goes." He was dressed in faded jeans and a T-shirt that boasted: *I Learned to Speak Hydrogen at the Wasserstoff Farm*. Over that he wore a curious white garment that looked like the illegitimate offspring of a lab coat and a toga. His feet were nestled into a pair of expensive-looking Nikes. He beamed with energy.

That made one of us.

I dressed, threw a little water on my face, and ran a comb through my hair. Zed stood by the open door, tapping an impatient foot on the stone floor.

"Don't s'pose I could get a cup of coffee?" I croaked, expectantly.

"There's coffee at the holosium," he told me, as we walked outside.

"At the *what?*"

"The holosium. It's right over there." He pointed to a large, circular building with a domed roof. The dome appeared to be a single, giant piece of smoked glass. Was it there yesterday? I had my doubts. "You can have coffee and eat your breakfast as we begin your lessons."

"Bacon and eggs?" I asked, hopefully.

He shook his head. "Fresh fruit, with cereal and soy milk. There are

no domestic animals here, unless..." He turned and smiled devilishly, "...you'd like to eat your horse?"

My horse? I'd forgotten all about poor Mike! All alone in a strange place, he'd surely wander off...

"Relax," Zed said, sensing my alarm. "He's right over there."

I spun around to see Mike grazing lazily by a pond, a portrait of contentment. He looked up as if to say, *Chill out, you silly uptight human; this is paradise!,* before refocusing his attention on the lush green grass by the edge of the pond.

Maybe I should take the cue from my horse, and quit trying to logically explain what was happening. Just go with the flow. I was, after all, in the perfect place to learn about hydrogen.

I took a deep breath. Then, filled with a new resolve to make the best of my time here, I asked Zed, "Why do you call this place the Wasserstoff Farm? I know 'Wasserstoff' is German for 'hydrogen', but, if you don't mind my saying so, Pickett is hardly a German name."

"True," he acknowledged, "but it's even less Greek. My ancestors roamed the forests of Northern Europe, along with the Celts and Germans. The Greeks—and the Romans, for that matter—called us all barbarians. Doubtless, we had a few choice names for them, too.

"But old rivalries aside, the English word 'water' comes from the Anglo-Saxon '*waeter*', which grew out of the Old High German '*wazzar*'. The German word, '*Stoff*', of course, means 'stuff', more or less, so

Wasserstoff is, literally, 'the stuff of water'. It's a good, unpretentious word that says what it means, and means what it says. Besides, it's got a nice ring to it, don't you think?"

Before I could answer, he continued, "The word 'hydrogen', on the other hand, is from the Greek words '*hydro*' and '*genesis*', and means '*substance giving rise to water*'. To my mind that's a bit more oblique than '*water stuff*'. It was coined by Antoine Lavoisier, a Frenchman who, when he wasn't naming things, was busy with important discoveries, such as the indestructibility of matter. He's the clever fellow who determined that matter is not actually destroyed when it is burned; in fact, all it really does is change form, an observation of no small import to our purposes here at the Wasserstoff Farm. Unfortunately for Antoine, he had also been a tax collector with the old regime at the time the French Revolution broke out. Try as he might to shed his former colors and become a good revolutionary—he was even on the commission that dreamed up the meter, the kilogram, and the liter—he was arrested in 1793, and later beheaded, on a fine May day in 1794. In response to a plea to spare his life, the Revolutionary Tribunal is said to have declared—a bit smugly, I'd say—that the Revolution had 'no need of genius'."

I looked up and saw we were at the holosium. Inside it was as dark as night. The room's only illumination came from a strange, glowing ball on a pedestal beside a table in the middle of the giant room. Zed motioned for me to sit in the table's single chair, where before me there was a vast assortment of cereals and tropical fruits. In the middle of the table a carafe of deliciously aromatic coffee rested on a burner with a small, barely-visible blue flame. A puddle of water collected in a saucer beneath.

"You eat," he instructed, bluntly. "I'll get started on your first lesson."

I poured a cup of coffee and bit into a mango as I waited for the show to begin. The mango tasted as if it had been picked this morning, and the coffee was several degrees fresher than any coffee I'd ever had.

I was about to compliment Zed on the wonderful cuisine he'd provided when, with a wave of his hand, a small, yellow dot appeared high in my field of view. There was no slide projector or computer console that I could see, nor did he hold a remote of any kind in his hand. I watched, mesmerized, as he spoke. "Behold, the birth of the universe!" He stabbed a bony finger at the yellow dot, and there was instantly an explosion of brilliant white light, followed by intense waves of energy. Suddenly the room felt very warm. I fought the temptation to duck under the table.

For the first few seconds all was chaos. Cascading streaks of color showered the room in wave after wave of glowing gas. Then slowly, almost imperceptibly, distinct rotating clouds of bright, hot gas began to coalesce from the snarl of light and energy. Soon, tiny bright specks appeared within the clouds until, after only a few moments, the dense clouds had all but disappeared, leaving billions of points of light, swirling like countless, distinct swarms of luminous gnats.

"Stars?" I asked, my head still spinning.

"Yes. And galaxies," he replied. "Or should I say proto-galaxies. This is about a billion years after the Big Bang; 12.7 billion years before the present, give or take. These stars are almost entirely hydrogen, with a little helium and lithium for good measure. At this point in the game there are no heavy elements like iron and gold, or even oxygen and carbon. They will all be created in succeeding generations of stars, formed from the remnants of these first stars, after their fuel is spent and they die in cataclysmic explosions."

I was impressed. Then a thought occurred. I asked, "But where did the hydrogen come from?"

Zed waved both hands in a circular motion and the scene suddenly changed. We were back at the beginning, though now, I sensed, on a submicroscopic level. The room filled with a hissing, roiling plasma, and out of it precipitated tiny blue and green spheres, bustling around and banging into each other like a multitude of blind kamikaze pilots. Fast as these spheres were, however, they were lethargic compared to the infinitesimal sparks flying across the room faster than the eye could follow.

"One second after the Big Bang," Zed announced, proud of his work. "This is unforgivably simplistic, I know, but it's the best I could come up with in so few dimensions. Anyway, the blue balls are the protons. They each have a positive charge. The green balls are neutrons, with no charge at all. Together they make up the nucleus of every element in the universe—except for hydrogen. The hydrogen nucleus is but a single proton."

With a wave of his hand everything disappeared but a single, spinning, blue sphere. "The hydrogen nucleus. Clean, simple, and doubtless not blue. But it's a nice visual all the same, don't you think?"

Zed sez

Wasserstoff is, literally, 'the stuff of water'. It's a good, unpretentious word that says what it means, and means what it says. Besides, it's got a nice ring to it, don't you think?

"So that's hydrogen?" I asked, "That little blue ball?" It seemed too simple.

"No, no, no!" He shook his head and his white locks swirled around his face. "That's a proton. It's the nucleus of hydrogen. For the rest, we have to speed through time."

I watched in amazement as the scene returned to the blue and green balls, moving ever slower. And the sparks I'd seen before, too fast for the eye to follow, now began to slow as well. After a few moments they started zooming closer to the protons, like comets nearing the sun, before spinning away again.

Zed pointed decisively at a single proton, and all the others vanished before my eyes. "This is 100,000 years after the Big Bang. The universe has now cooled to just a few thousand degrees." One of the many sparks that had been zooming about the room at breakneck speed now settled into an orbit from a great distance around the blue proton. In fact, it orbited so fast the proton was all but obscured by the cloud left in its wake.

"Now *that's* hydrogen," He announced, "one positive charge, happily coupled to one negative charge—though hardly to scale."

"What do you mean, 'hardly to scale'?" I asked.

"The room's not big enough for scale."

"Well how big would it have to be?" I wondered aloud.

He pierced me with an exasperated glance, then waved his hands wildly. The Earth appeared before my eyes. "Recognize this? It's Earth. Now watch." The globe turned slowly until North America came into view. Then a tiny piece was excised, by what looked like a laser beam, from just east of the Rocky Mountains and sent toward the edge of the circular room, where it began spinning fervently around the Earth. Two new globes popped into existence and fell into place. One began to orbit between Earth and the piece that had been removed from it. This was a small red planet, Mars. An equal distance on the other side of the orbiting city—and almost to the edge of the large domed building—there appeared a calico planet, Jupiter. "If the proton—the nucleus of hydrogen—were the Earth, then the electron would be Denver, circling the Earth from about the midpoint between the orbits of Mars and Jupiter."

I'd been to Denver enough to know that that wasn't quite far enough away to please me, but it was a good start.

"Or," he said, "if that's too much of a reach for your imagination, try this." With a spin of his finger the Earth became a giant beach ball. Denver, Mars and Jupiter blinked out and disappeared. Behind the beach ball a breathtaking panorama appeared; snowcapped mountain peaks, lined up one after the other in the distance.

Zed spoke: "In this scenario the proton is the beach ball; a really big beach ball. The electron is a teeny, tiny grain of sand orbiting the beach ball from 10 miles away." A grain of sand appeared before my eyes, then sped off toward a distant mountain, where it began to glow brightly. Then it commenced to spin about the room, and everything within, at impossible speeds. Soon we were within a dense, opaque cloud created by a single grain of sand spinning about an oversized beach ball from far beyond the confines of the room. I couldn't see Zed, or even the slice of pineapple I was attempting to stuff into my mouth. I felt like I was trapped within a magnetic field of infinite intensity. My hair stood on end and a tingling sensation invaded every cell of my body. My right half was drawn one way, my left half the other. I wanted to scream, but my vocal cords were working against one another. It was a nightmare. I wanted to run from the room, but the entrance was obscured by the dense electron cloud. Not that it mattered: I was pinned to my chair. Then, just when I thought I was about to pass out from the polarization occurring in the cells of my brain, I heard Zed scream, "ENOUGH!"

And everything returned to normal.

I looked at my watch in the light of the globe beside the table. It was 9:00 a.m. My nervous system had lived through an entire day, and I hadn't even finished my breakfast.

"Any more questions?" Zed asked.

From somewhere in the foggy recesses of my mind, I heard myself mumble, "Got any donuts?"

technistoff - 2

Zed's colorful dramatization of the birth of the universe (AKA the Big Bang) and the creation of the first elements was just about the best lightshow I'd ever seen. But was it accurate? Not knowing Zed all that well on that second day, I thought I should try to verify the results of his wizardry. I was happy to find that his version of the Beginning was remarkably close to the account given by the eminent physicist, Stephen Hawking in *The Universe in a Nutshell*. What was missing from Hawking's account, of course, was the vivid animation. Zed most certainly took creative license with the sound effects and the colors, but I can't begrudge him that. As I suspected at this point, and verified later, Zed is at least as much a showman as he is a teacher and wizard.

protons & electrons

His analogies for the comparative sizes of the proton and the electron, and the relative distance between them, were a little harder to verify. Of course, the fact that Zed never reveals his sources—preferring, instead, to act as though it's all fact written in some cosmic record that only he can read—makes it even harder.

Be that as it may, any good physics or chemistry book will give the diameter of the proton at 2×10^{-15} meters. The size of the electron is a little harder to nail down, but the generally accepted absolute upper limit is 2×10^{-18} meters, or 1,000 times smaller than the proton. (Most authorities feel the electron is much smaller, and many feel it has no size, at all, though no one knows how to prove it.) So, if the beach ball were 2 feet in diameter [about 0.60 meters], then the grain of sand would have to be about 0.60 millimeters across. As Zed pointed out, that's a very, very tiny grain of sand. By figuring the city of Denver (minus the 'burbs) as 8 miles *[12.87 km]* across, it would be 1000th of the diameter of the Earth. (Denver may be bigger than that, but I liked the concept so much I was reluctant to insist Zed use a different city.)

That leaves us with the distance between the proton and the electron. This is given by an equation known as the Bohr Radius, and is equal to 0.529×10^{-10} meters for the innermost permissible electron orbit. By dividing the Bohr Radius by the proton's diameter, I concluded that the grain of sand was 26,450 beach ball diameters away. If you take 26,450 times 2 feet, you will see it equals 52,900 feet, or 10.02 miles *[16 km]*.

In the same way, 26,450 times 8,000 miles puts Denver in orbit 304,600,000 miles *[490,100,000 km]* from the sun (211,000,000 miles past Earth), which is about the midway point between the orbits of Mars and Jupiter.

The Bohr Radius should be in any good physics or chemistry text. For the comparative sizes of subatomic particles, a wealth of interesting information can be found at: *http://particleadventure.org/particleadventure/index.html.*

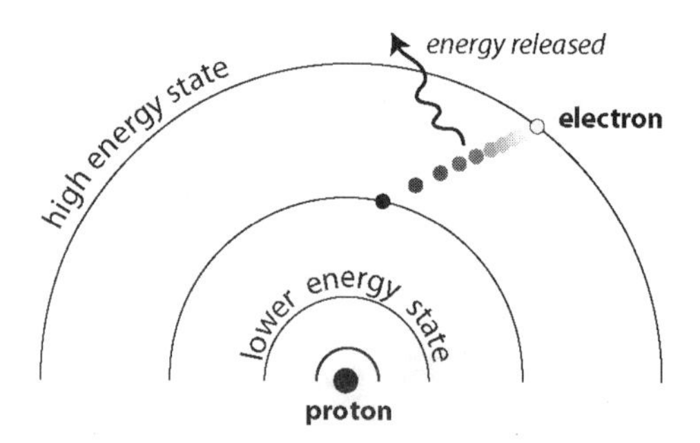

Hydrogen Energy States

Energy is released when the hydrogen electron moves from a high energy state (outer orbit) to a lower energy state (inner orbit)

BY THE WAY: What would *really* happen if you had the bad luck to find yourself stuck between a proton and an electron? The fact that it's an impossible question notwithstanding, it's safe to say that Zed's little demonstration was unrealistically innocuous.

Not that I'm complaining.

chapter 3

Unearthly Reactions: Atoms vs. Atomics

I was hoping for at least a brief respite before resuming my lessons. A pleasant horseback ride around the grounds would have gone a long way toward rejuvenating my battered brain. But Zed was just getting started. He studied me in an uncomfortably curious way—as a cat would study a mouse, I thought—until he was certain I was ready for more. Then he continued.

"That last demonstration, you must understand, was hydrogen's lowest energy state, what physicists call the 'ground state'. There are three more possible orbits for the grain of sand around the beach ball, each farther out and more energetic than the one before. As the electron cascades down through the various orbitals, it gives off energy."

A question occurred to me, so I asked, "Why doesn't the electron spin into the proton? I mean, negative attracts positive, and vice versa, right? The orbits of satellites decay, right? Like Skylab did several years ago? What stops electrons from doing the same thing?"

Zed smiled broadly, and answered, "An excellent question, lad! Glad to see you're thinking. But before I give you an answer, you need to understand that an electron is really nothing like that grain of sand. The truth is, it's more like a waveform than a distinct particle. In fact, many physicist believe an electron has no extension in space, whatsoever; that it may be what they call a 'point-like particle'. But that's neither here nor there. The important thing is this: the fact that the electron moves with a wave-like motion means that its journeys around the nucleus are confined to discrete orbits—it has to end up in the same phase after each revolution;

a trough has to line up with a trough, a crest with a crest. It's like drawing a circle with a wavy line; you have to draw it just the right circumference, or you won't end up where you started. Likewise with electrons; if the orbits were shifted to any degree at all, it would throw everything out of kilter. The world would be utter chaos and we wouldn't be having this conversation."

"No, I s'pose not."

Path of an Electron

A simplistic illustration of the wavelike nature of an electron's orbit

"Anyway," he continued, "it's time to muddy the waters a little, so to speak." He pointed toward the darkness of the dome and the familiar spinning blue sphere appeared. A tiny point of light appeared a short distance away and began rotating around the sphere in a slow, wavelike orbit. It was not at all to the scale of the beach ball and the grain of sand. I was ready to duck under the table in the event things started getting out of hand again, but the display remained small, slow, and unthreatening. "Ordinary hydrogen," he announced. "One proton, one electron. As simple as it gets." Using his finger as a pointer, a green ball precipitated out of the ether and came to rest snugly against the blue ball. The two balls began spinning together as one, making the nucleus appear blue-green. The orbiting electron seemed unperturbed. "Now we have deuterium; heavy hydrogen. A nucleus of one proton and one neutron. It's what makes "heavy" water heavy. It's used mainly to slow down high-energy neutrons in atomic reactors, to help the reactions along. Roughly one in every 7,000 hydrogen atoms in water is deuterium, so for every gallon of water you drink, you're drinking a little over ½ gram of heavy water."

"So how heavy is heavy water?" I asked, taking a sip of coffee.

"About 10 percent more than ordinary water, is all. A quart weighs in at about 35.5 ounces, or if you like, a liter is about 1.1 kilograms. That's because most of the weight of water comes from the oxygen atom."

He paused to see if I was finished with my petty questions, then pulled another green neutron out of thin air and cast it into the spinning deuterium nucleus. The new nucleus grew in size and changed color from blue-green to green-blue. Still the electron was unfazed. "This is tritium, the third and final isotope of hydrogen..."

"What's an isotope?" I piped in.

He pierced me with an impatient stare, then said, "Isn't it obvious? It's a form of an atom with the same number of protons and electrons—one each, in the case of hydrogen—but differing numbers of neutrons."

"Oh, sure," I said, feeling embarrassed.

"Anyway, this stuff is exceedingly rare. And it's also radioactive. That means the nucleus is unstable and has a tendency to cast out one of the two neutrons." With an audible *pop!* a neutron was ejected from the new nucleus and zoomed across the room, careening off the wall behind me. "This makes it important in thermonuclear reactions..." I started to open my mouth to speak, but he cut me off, saying, "...a subject we'll get to shortly."

"So this is it," he continued, "the building blocks of matter. With enough protons, neutrons and electrons you can make any element you want—from helium to plutonium; provided, of course, you have either god-like powers, or ready access to obscene amounts of energy. But, since you and I have neither, we'll have to be content with my crude holographic representations."

He gave me a hard glance. I just smiled, and munched on a banana.

"Okay, then. Shortly we'll get to the elements of life which, as I'm sure you know, are also the building blocks of what we call fossil fuels—

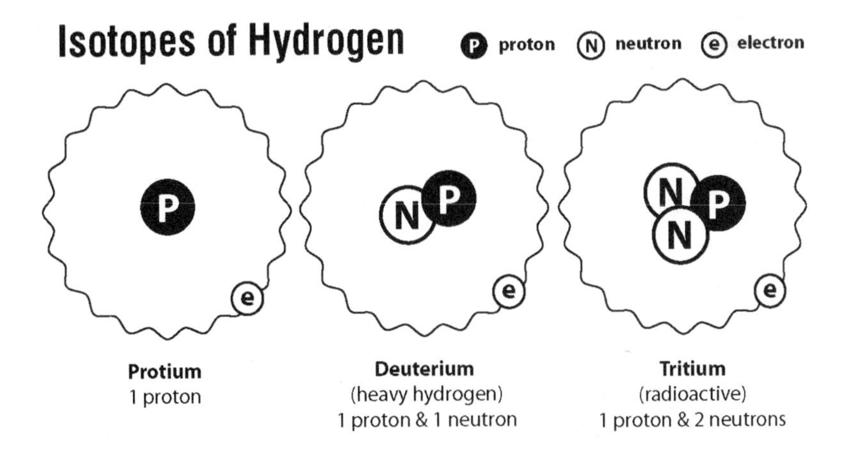

the whole reason the planet's in such higgledy-piggledy and why this hydrogen thing is getting so much attention. But first we've got one more element to cover, namely helium." With a slash of his finger through the air, the deuterium nucleus stopped spinning, and the electron froze in space. Another deuterium atom appeared beside it. Then the two nuclei merged and the extra electron moved opposite the first. With a twist of his hand the new nucleus began to spin and the electrons resumed their wavy orbits around it. "This is helium, or at least the most prevalent isotope of helium. It's the happiest element in the universe. It has everything it needs and has no desire to react with anything."

I hated to interrupt, but I had to ask, "What do you mean, 'everything it needs'?"

"Two electrons balanced by two protons, balanced by two neutrons. It's quite content to ignore the rest of nature," he assured me.

The Helium Atom

2 electrons spin in an orbit around 2 protons and 2 neutrons, making helium the happiest (most chemically inert) element in the universe.

"But what about deuterium?" I asked. "Doesn't it have everything *it* needs?"

"Not quite. Notice for helium the two electrons spinning nicely together in the same orbit? That orbit is called an energy level. There are several energy levels in heavier elements, each capable of holding more electrons than the one below it. The first energy level is full with two electrons; the next needs eight, and so on. In our helium atom, the positive charges of the two protons are balanced by the negative charges of the two electrons. That energy level is full."

I was catching on. I said, "So, even though hydrogen has a proton and an electron, and therefore no net charge, the electron still wants another electron?"

"Exactly! That's why hydrogen is so reactive."

Zed sez

With enough protons, neutrons and electrons you can make any element you want—from helium to plutonium; provided, of course, you have either godlike powers, or ready access to obscene amounts of energy.

I pointed to the animated helium atom, and said, "That looked pretty easy. Why doesn't hydrogen just turn into helium? Wouldn't it be 'happier', as you say?"

Zed shook his head, then pulled his long, white locks away from his face. "Things don't work that way. Electrons are spritely things, and not too long on loyalties. They can come and go on a whim. But nuclei are stalwart animals. They're bound together most tenaciously by what is aptly called the 'strong nuclear force'. There's hell to pay when you start ripping atomic nuclei apart."

I was beginning to catch his drift—Hiroshima, and the like. I asked, "Okay, then what happens in the sun? Doesn't hydrogen turn into helium?" I really hoped I wasn't going to regret asking that question.

He smiled menacingly, and replied, "Oh, indeed it does, but not like our last simple—but erroneous—demonstration." With a wave of his hand, the happy, benign helium atom disappeared, replaced by four ominous blue protons, spread out like the corners of a square. I was starting to cringe.

"The sun produces all of it's heat and light—which leads to all life on Earth—by taking four protons and fusing them together..." I watched with dread as the protons each made a beeline toward the center of the square. When the inevitable clash occurred there was an intense, though soundless, explosion of light and heat. The holosium lit up like the inside of a million-watt bulb and, for a brief moment, felt like a toaster oven set to 'cremate'. I buried my face in my hands until the spectacle subsided. Zed was unperturbed. "...into a nucleus of two protons and two neutrons. The helium nucleus, you see, weighs just a tad less than the four protons, and that little bit of extra matter is converted into energy by way of Einstein's equation, $E=MC^2$."

"Is that what happens inside nuclear reactors—and please, Zed, no demonstrations?" I was quick to add.

He looked terribly disappointed. "No, just the opposite, in fact. In nuclear reactors a heavy, unstable element—like uranium 235—is broken down by neutron bombardment into lighter elements. The heat of the reaction changes water into steam, which turns huge electric turbines. Unfortunately, the byproducts are highly radioactive, which means they tend to spit high-energy neutrons out of their nuclei in alarming numbers. High-energy neutrons, as you might have guessed, can be really hazardous to your health.

"Those are fission reactors; hasn't been one of *those* built in the U.S. for many years. Three Mile Island and Chernobyl really left the public feeling bitter and betrayed.

"But that's not the end of it," he went on. "There's way too much power in the atom to give up on it. The holy grail of cheap, clean energy, of course, is the *fusion* reactor, which works pretty much the same way the sun does. These plants will be much safer, if they ever manage to build one, and they will produce no long-lived radioisotopes. Unfortunately, they don't exist, yet. The poor physicists working on the problem just can't seem to produce enough heat to sustain a reaction, or a proper vessel to contain it. At least not on an economically viable scale."

I had one more question, though I was loath to ask it. I fidgeted in my chair, unable to make up my mind, until Zed took notice of my indecision and said, in his silvery-tongued way, "Spit it out, lad. Don't choke on it. What is it you want to know?"

"Well," I started, with due hesitancy, "I was just wondering: what is a hydrogen—or should I say, a *thermonuclear*—bomb? Is it like the sun, where a bunch of protons get all bunched together into helium nuclei? And if that's really the way it works, why are thermonuclear bombs so radioactive?"

Zed's eyes flashed darkly, and he offered a grim smile. "A thermonuclear bomb? Now *there's* a nasty piece of work!"

"Can you explain it without blistering my skin?" I asked, hopefully.

"Of course, lad," he said consolingly, "of course I can."

He scratched his chin and thought for a moment, before answering. Finally, he began. "You see, it's not the thermonuclear explosion that's so radioactive; it's what you have to do to create the heat to make the thermonuclear explosion." His fingers started to twitch and I started to worry.

"To create a fusion reaction," he began, as he made several small sticks marked *TNT* appear in front of me, "you have to create a fission reaction. And to do that you have to ignite a lot of high explosives; TNT works fine." The individual sticks of TNT suddenly fused together into a membrane enclosed within a transparent bomb casing, and wrapped itself around a glowing and ominous green mass.

"The idea is to direct the power of the TNT implosion to compress a handful of radioactive uranium, U-235—or better yet, something really reactive, like plutonium, Pu-239—to the point that it begins to emit neutrons at a prodigious rate. The liberated neutrons careen into other nuclei, liberating more neutrons, until an exceedingly fast chain reaction occurs."

Below the fission bomb, but still within the confines of the casing, a heavy-sided cylinder appeared. At the center of the cylinder there was a greenish rod.

"This is bit simplistic, I'm afraid," he said, pointing to the new additions to his hologram, "but it will have to do."

"It doesn't look at all simple to me," I remarked.

"You see, the rod in the center is made from plutonium-239. Lithium deuterate fills the spaces around it. That's a compound made from deuterium and lithium, the latter being useful in the production of tritium, which, as you should remember, is radioactive hydrogen. It's handy for the production of free neutrons. Understand?"

"Sure." I was listening, but my eyes were focused on the bomb suspended just below the ceiling of the holosium. I watched as the TNT shroud silently imploded, and the green mass within glowed ever brighter. So far, Zed was being nice.

Zed sez

There's way too much power in the atom to give up on it. The holy grail of cheap, clean energy, of course, is the fusion reactor, which works pretty much the same way the sun does.

"The fission explosion does two things: first, it releases massive amounts of free neutrons and x-rays, and second, it creates intense heat—even more heat than is produced inside the sun. That's why you get an explosion and not simply a long, sustained reaction, as takes place in the sun."

The green glob of holographic uranium suddenly transformed into the light of the sun, though, true to his word, Zed hadn't yet boiled my hide.

"At this point," he continued, "the heat of the atomic blast is directed downward, toward the lithium deuterate."

"What about all the electrons?" I wondered.

Zed roared with laughter. "Electrons? Lad, they headed for the high ground a long time ago. All we're dealing with here are nuclei!"

Of course. How stupid of me.

"Anyway," he went on, "a compression shock is initiated within the lithium deuterate. This compresses the plutonium rod into a supercritical mass. This mass, in turn, detonates, giving off intense heat and radiation, which causes a fusion reaction within the lithium deuterate. As you can imagine, all the heat, x-rays, and high-energy particles flying around make for a devilish mix of reactions. In the end, the universe has a trace more helium, a little less deuterium and lithium, and the Earth has a radioactive scar that will take eons to heal."

As he spoke I was watching the bomb casing behind Zed's head. He must have forgotten about it, because it seemed to be going super-critical. I screamed, "Zed!" but it was too late. I ducked under the table as Zed looked at his creation with an equal mixture of pride and horror. The blast was ear-shattering, and the light it produced was like a thousand suns. A shock wave followed, blowing the table above me into the wall like a paper prop. I remember thinking, *there goes breakfast!* as I rolled into the fetal position and held my hands over my ears. Heat followed the shock wave. I felt as though I'd been shot out of a geyser, into a volcano.

After a few moments, the heat and light subsided. I looked up. Zed was still standing, looking into the glowing embers of his holographic catastrophe. He turned to face me. His face was red and his ivory hair a little singed, but, amazingly, not much worse for wear. He smiled with an otherworldly detachment. "I trust we've reached the end of this lesson?" he said.

Indeed we had.

technistoff - 3

If Zed was hoping to impress me with this lesson, he overshot his mark by about three orders of magnitude. As far as I could tell, he was trying to draw a clear distinction between the two processes for getting energy out of the hydrogen atom: 1) between the energy given off when the hydrogen nucleus is fused into a helium nucleus; and 2) when—in a vastly more benign reaction—the molecular bonds (which is to say, the electron bonds) are broken in the diatomic hydrogen molecule (H_2). I'd have been happy with a few scribbles on a chalkboard, thank you, Zed.

But, I sooned learned, Zed just does what Zed does best.

atomic energy

However scary Zed made it seem, hydrogen fusion technology could be the promise of the future. If anyone could actually figure out how to pull it off, we could have clean, safe and virtually limitless energy. But for now we will have to content ourselves with the comparatively feeble energy that's given up when the hydrogen molecule is broken.

I was able to verify Zed's claim that the ratio of normal water to heavy water is 1:7000 by referring to *Chemical Elements and their Compounds*. The weight of heavy water was easy to calculate from the atomic weights of normal hydrogen (1.00794) and oxygen (15.9994) by simply doubling the value for hydrogen. It works to within a range that's good enough for anything short of a technical journal, since a neutron's mass is nearly identical to a proton's, and an electron's mass is too small to make a noticeable difference.

The fusion of hydrogen to helium in the sun is described by innumerable sources. It first caught my eye in the May 2001 issue of *Scientific American*, in an

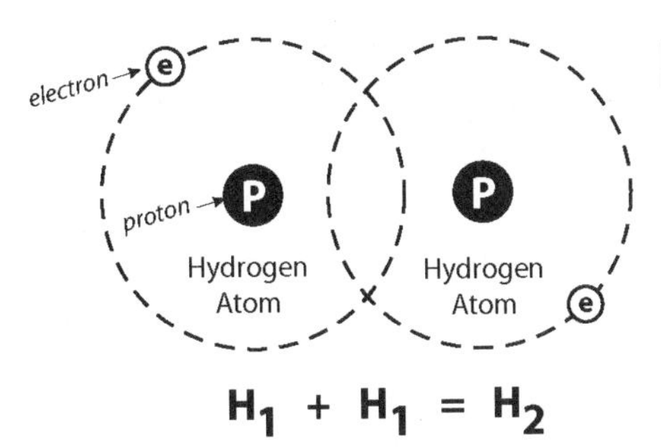

Diatomic Hydrogen Molecule

Covalent bonding of 2 hydrogen atoms means that the 2 electrons are shared, and each atom now has 2 electrons.

article by Brian C. Chaboyer, titled *Rip Van Twinkle*. You may be baffled when you read this article, since the author states that four protons outweigh a helium nucleus of two protons and two neutrons (by about 0.7 percent). It's confusing because a quick glance at any physics book will show you that neutrons are slightly heavier than protons, so you would naturally think the helium nucleus should be *heavier*, not *lighter* than four protons.

This discrepancy is easily resolved without too much hair pulling once the bonding energy of the nucleus is taken into account. When protons and neutrons combine in a nucleus, part of their mass becomes the energy that holds the nucleus together. It's the $E=MC^2$ thing, again, where energy is equal to mass, times the speed of light squared. In other words, there's a lot of energy in a little bit of mass, and that's what makes nuclear reactions so devilish.

There are many recipes for making fission and fusion weapons. The internet abounds with them. For a thoroughly readable account of the technical and theoretical difficulties that had to be overcome to produce the first atomic (fission) bombs, read Richard Rhodes' *The Making of the Atomic Bomb*. For a really good story wrapped around a detailed account of the difficulties involved in making a hydrogen bomb...excuse me, a thermonuclear device...read Tom Clancy's *The Sum of All Fears*. If you'd like to detonate a Teller-Ulam fusion bomb in cyberspace, visit *http://science.howstuffworks.com/nuclear-bomb7.htm*.

It's a blast.

Fusion vs. Fission

Atomic **fusion** occurs when light nuclei—such as four protons, or hydrogen nuclei—unite into a heavier nucleus—such as a helium nucleus, with two protons and two neutrons. A fraction of the atomic matter is converted into a tremendous amount of energy.

Atomic **fission** occurs when a heavy nucleus breaks apart, usually because of bombardment by neutrons, into two lighter elements. As with fusion, a fraction of the atomic matter is converted into a very large amount energy.

chapter 4

Chemical Bonds: The Numbers Game

I'd had enough. In less than two hours I'd been magnetically seized inside the orbit of an electron, nearly fried—twice—and subjected to enough heat, pressure and wind to last a lifetime.

When I was certain I could move, I jumped to my feet and ran for the door. Zed was right behind me. "Wait," he insisted, "we're not finished yet!"

I ignored him. Hurrying through the door, I didn't bother to slow down until the holosium was a speck in the distance. I needed to be outside with nature for awhile. I saw Mike up ahead. My heart leapt with joy. I ran to him and jumped on his back, no bridle, no saddle. Wherever he wanted to go, and however fast he wanted to get there, was fine by me.

Mike pivoted toward the inside of the Wasserstoff Farm, and headed serenely into the interior. He was in no hurry at all. He stopped at every rock outcropping, every growth of trees, to sniff and look around. He even took time to smell the flowers.

Flowers?

It was early November in Colorado. The flowers should've been a sweet but withered memory. Here they bloomed in abundance. And the grass! It was verdant green. Why hadn't I noticed it before? My eyes turned to the broadleaf trees: aspen, cottonwood, willow and scrub oak. They were all leafed-out in a mid-summer sort of way.

And why not? It was warm; paradise should be so nice. Except for a steady breeze, everything was perfect.

How was this happening?

Suddenly, I wasn't done with Zedediah Pickett and the Wasserstoff Farm. I needed to know more.

No sooner had this thought crossed my mind than Mike jumped in response to the sound of rustling leaves behind us. I grabbed his mane as he spun around to face—who else?—Zedediah Pickett. Looking as composed as ever, he stepped out from behind a small stand of aspens. We must have been half a mile away from the holosium. How had he gotten here so fast?

"Feeling better?" he asked. If I was hoping to hear a note of contrition, I was disappointed.

"A little," I replied.

"Sometimes demonstrations are the best way to make a point."

"As long as they don't get out of hand," I was quick to add.

"Point taken. But admit it—you'll never have any doubts about how messy things can get when you start rending atomic nuclei."

"Boy, howdy."

"And there was no harm done, was there?"

"No, I s'pose not," I grudgingly agreed, "at least as long as holographic radioactive fallout isn't dangerous."

"It's perfectly illusory," he assured me.

"Glad to hear it, 'cause that illusion of yours plastered my breakfast all over your nice, clean wall."

He smiled, proudly. "It was a good one, wasn't it?"

"A little too good, maybe?"

He sighed as though he were tiring of the conversation. He said, "However you choose to look at it, it's in the past, while all your remaining lessons are still in the future. So let's get busy, shall we?"

Why not? I slid off Mike's back. The formerly fractious horse was content to stand placidly beside us. "I just hope this isn't a lesson that requires the holosium, because I've had my fill of that place for awhile."

"Oh, I think we can manage without it. In fact, we should be able to do this lesson with balls and sticks."

"You don't say?" It sounded innocuous enough, but I was still leery.

He waved his hands in that spooky way of his, and a pair of small, glass spheres materialized before my eyes, along with a pencil-sized glass rod. Zed instructed me to hold out my hands. When I did, his "illusory" toys fell into them. I was so surprised I nearly dropped them. No wonder he hadn't needed a projector or remote control device inside the holosium; it was all mind over matter for Zed.

"There you have it—two hydrogen atoms. See what you can do with them."

The spheres vibrated with energy. I turned them over in my hands and discovered each sphere had one hole in it. I stuck a sphere onto each end of the glass rod, making what appeared to be a transparent, leprechaun-sized dumbbell. I said, "How exciting. What's it good for?"

"Nothing, yet. What you hold in your hand is two hydrogen atoms as they like to be...a hydrogen molecule. The rod represents the bond that holds them together—the two shared electrons. With two protons in the nucleus and two electrons orbiting both of them, the first electron energy level is full. And, since the positive charge of the two protons is equal to the negative charge of the two electrons, the diatomic molecule has no net electric charge. It's as content as two lonely hydrogen atoms can be, though, the truth be told, hydrogen by itself is never really all that happy."

"Oh? Why is that?" I asked, thinking it was a strange thing for Zed to say.

He answered, "Because hydrogen-to-hydrogen bonds are weak, just by their nature. Hydrogen feels much more useful when it's attached to an atom different from itself." He motioned in the air with his index finger, and said, "Here, catch!"

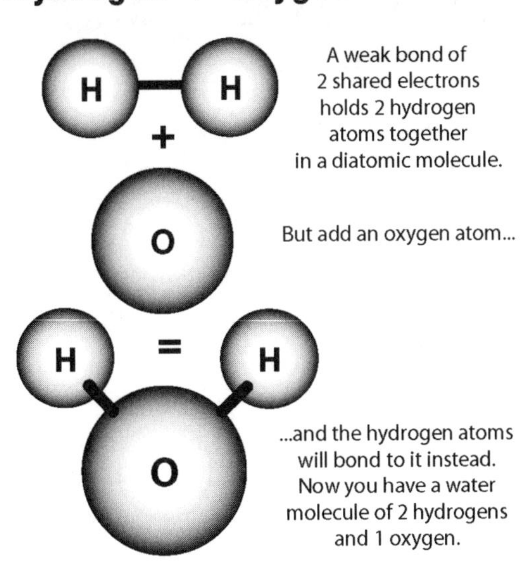

Another glass sphere appeared—much larger than the other two—along with another glass rod. I held out my hand just in time.

Zed said, "There. See what you can do with these."

Zed sez

Along comes a pair of hydrogen atoms, with one electron each. 'Care to share?' oxygen asks. 'Sure, why not?' hydrogen replies. Then boom! Water. And everyone's happy.

I turned the big sphere around in my hand. It felt full of energy; much more so than the smaller spheres. I noticed that it had two holes, instead of just one. They were located a little less than a third of the circumference away from one another. I rearranged the rods and spheres until I held in my hand what looked like a featureless bug's head with a little ball on the tip of each of its short antennae. "Cool," I remarked. "What is it?"

He blinked his eyes and twitched his nose. The bug's head quickly melted, and my hands were suddenly dripping wet. "Water?"

Zed nodded. "Good old H_2O—two atoms of hydrogen and one atom of oxygen, bound very tightly together. It's a good arrangement."

"I'm sure, but why is the oxygen bound so tightly to the hydrogen?"

He speared me impulsively with that impatient look of his, then suddenly smiled. "Because, lad, the hydrogen is exactly what the oxygen wants. Oxygen has eight of everything: 8 protons, 8 neutrons, 8 electrons. Sounds nice and symmetrical, doesn't it? But, if you remember our last lesson, you will recall that electrons exist in distinct energy levels. The first level holds two, and the second holds...?"

"Eight?"

"Correct! With two electrons filling up the first energy level, the second one only has six. Along comes a pair of hydrogen atoms, with one electron each. 'Care to share?' oxygen asks. 'Sure, why not?' hydrogen replies. Then *boom*! Water. And everyone's happy. All the energy levels are filled and the atom has no net charge."

I was a little confused. "You mean those two hydrogen electrons orbit around *both* hydrogen atoms *and* the oxygen atom? Sounds like a wreck waiting to happen."

Zed chuckled. "Not really. Remember the beach ball and grain of sand? There's a lot of space in there. Besides, they've been doing this for billions of years; plenty of time to work out the choreography. The fact is, it's even more complicated than what I've just told you. Have you ever wondered why carbon dioxide is a gas at room temperature, while water is a liquid, even though CO_2 is over twice as heavy?"

"Not until now," I admitted.

"Well, it all has to do with the hydrogen bond. Certain elements—oxygen, primarily, but also nitrogen and fluorine—want hydrogen's electron so badly they pull it into an uneven orbit. This leaves hydrogen's positively charged backside blowing in the breeze, so to speak. It turns water into what is called a dipolar molecule: a positively charged end and a negatively charged end. The positively charged end attracts the electrons from oxygen atoms in nearby molecules like ambulances attract lawyers. The end result is that every water molecule is drawn toward every other water molecule. Water is more than just a polygamous marriage, it's also a commune; something like a three-dimensional chain, or a forever changing crystal. A fortunate arrangement, to say the least. Without hydrogen bonds, all the ice on Earth would melt into water, and all the water would soon evaporate."

"Sounds dreadful," I opined.

"To say the least."

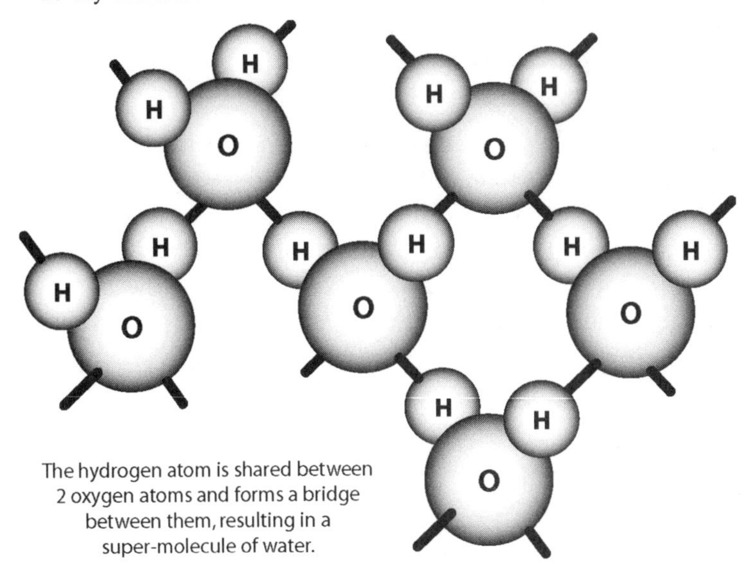

Hydrogen Bridge Between Water Molecules

"Hey, wait a minute!" I objected. "I thought you said earlier that hydrogen bonds weren't all that strong!"

Zed shook his head. "No, I said hydrogen to hydrogen bonds weren't all that strong. I wasn't talking about *the* hydrogen bond, between hydrogen and oxygen. See the difference?

"Yea, okay." Then another question popped into my mind. "So what else can you do with hydrogen and oxygen?" I asked.

"Well, you can add another oxygen and make H_2O_2, otherwise known as hydrogen peroxide. That extra oxygen doesn't really fit in, however, so it doesn't hang around any longer than it has to. That's why hydrogen peroxide always comes in a brown bottle—sunlight very quickly breaks it down into water and oxygen gas. So does blood; that fizzing you see when you pour it on a cut is the extra oxygen escaping back into the atmosphere."

It was all interesting, but I was getting frustrated. "This is going nowhere," I complained. "We need another atom to liven things up."

Zed smiled broadly. "Indeed we do, my young friend. Would you care to make a suggestion?"

I thought for a moment, then said, "Okay, how about carbon?"

"An excellent choice! Why, with just carbon and hydrogen you can make dozens of compounds. Methane, butane, propane, acetylene and naphthalene are all just carbon and hydrogen. And if you add in a little oxygen, you hit the chemical lottery."

"Good! What do I win?" I wondered.

"How about life itself? With carbon, oxygen and hydrogen you can make all the sugars, fats and carbohydrates in the world!" His fingers danced wildly in the air, like he was playing an invisible keyboard. A bizarre menagerie of chemical creatures paraded in front of my eyes. Some resembled centipedes, others were coiled up like snakes. Still others had shapes beyond my meager powers of description. There were hundreds, no, thousands of them. I was impressed.

"How about protein?" I wondered.

"Well, if you'd allow me to throw in a teensy bit of nitrogen and sulfur, I think we could manage that too."

"If you must," I sighed.

Zed went back to work. Now it was all snake-like chains of atoms with little legs poking out all over, all twisting and writhing as though there was no way to get comfortable. Soon the chains of protein began to mingle with the previous molecules. Then the whole show shrank in size as more molecules materialized and joined the fun. Soon, the tangle of molecules began to coalesce into recognizable tissues: hair and skin and muscle. "Now for a little calcium and phosphorus," he said, never taking his eyes off his work. Bones began to form in front of my eyes, and muscles, skin and hair wrapped around them.

Zed was on a roll. His eyes blazed with an inner light. Everything was self-assembling and changing in scale so quickly it became a blur. After several moments I thought I could discern a recognizable shape appearing out of the chaos. A moment later, a curious brown house mouse hung suspended in the air in front of me. "I had to throw in a smidgen of a few dozen other elements, when you weren't looking," he admitted.

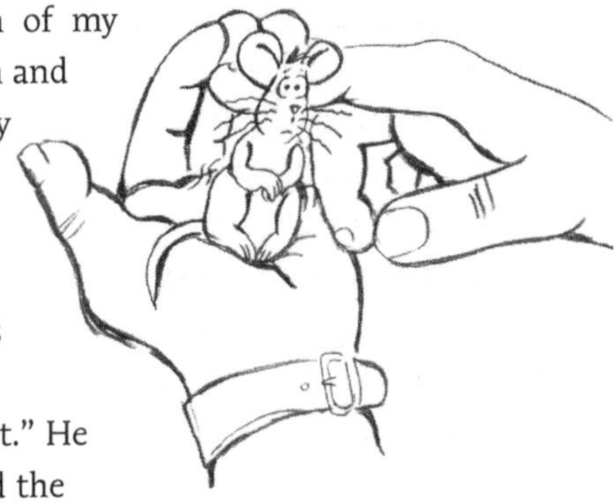

I reached out with the palm of my hand and picked it up. It was warm and soft. It looked up at me expectantly with beady, nervous little eyes. I couldn't help but run my finger along its back. It felt like a mouse. "Zed! Is it...is it real?" I was unable to stifle my wonder.

"What, that? No, of course not." He moved his palm over my hand and the mouse disappeared. "It takes a mouse to make a mouse. Do I look like a mouse to you?"

"Well, no...but..." I was sad to see the mouse vanish so quickly.

"I was just making a point. Why do you always get emotional whenever I make a point?"

"Because," I observed, "the last time you made a point I almost got skinned alive."

"Oh, stop with that already. It's over! Anyway, we seem to have veered off course here. One minute we were adding carbon to hydrogen and oxygen, and the next minute we're shaking in spice and making mice. Heavens, but I do get sidetracked!"

Trying to be helpful, I said, "Okay. We'd reached a dead-end with hydrogen and oxygen, so you decided to add carbon and things suddenly got out of hand. But you never did explain why carbon makes all these complex molecules possible."

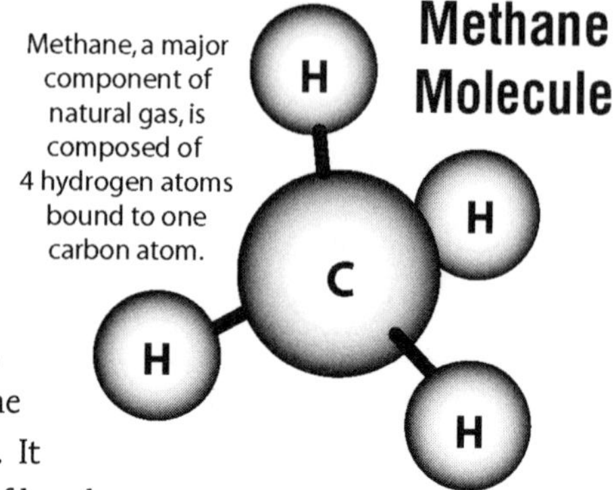

"Right. Of course," he said, getting his bearings again. "The reason is not terribly complex. It all has to do with the number of bonds carbon is naturally inclined to form with neighboring atoms. If you remember, hydrogen forms a single bond, oxygen forms two. We ran out of compounds rather quickly."

"Yes, we did," I agreed.

"Well, carbon, having only four electrons in its outer energy level, likes to form four bonds with its neighbors. It doesn't take a lot of imagination to see the infinite possibilities. That's why 93 percent of the mass of humans...and, uh, mice...is hydrogen, oxygen, and carbon. Carbon provides the matrix for hydrogen and oxygen to cling to. These three atoms alone make up practically everything plants and animals consider as fuel. And since coal and oil and natural gas are all derived from living tissues—mostly plant tissues, of course—most of what we consider fuel is made from some combination of hydrogen, oxygen and carbon."

"Why just those three?" I asked.

"Good question. Nature likes to use what She has at hand. Hydrogen and oxygen are constituents of both water and air. Besides that, they're light and readily available, so they were chosen."

"Wait a minute!" I objected. "Isn't there something that likes to form 3 bonds?"

"Sure. Nitrogen. It weighs in right between carbon and oxygen. It needs 3 electrons to fill its outer energy level."

"You'd think it would be a bigger player in the game of life, wouldn't you?"

Zed looked at me indignantly, and replied, "You might. I wouldn't."

Exasperated, I asked, "Why?"

"Availability," he answered.

"What do you mean, 'availability'? The atmosphere is full of nitrogen. It's 79 percent nitrogen, if I recall."

"Indeed it is," Zed agreed, "but how much does your body utilize every time you take a breath? Zero, that's how much. Oxygen, on the other hand, is less than 20 percent of the air you breathe, yet you would be dead in less than 5 minutes unless your lungs could use that oxygen. Why do you suppose that is?"

I didn't have a clue, and it was written all over my face.

Zed smiled kindly, having bested me again. "Nitrogen, like oxygen and hydrogen, exists in nature as a diatomic molecule: one atom of nitrogen bound to another. The difference is the nature of the bond. Hydrogen-hydrogen bonds, as you already know, are easily broken. Same with oxygen. But nitrogen? It's extremely fond of itself, and it forms very high energy bonds. It's no accident that gunpowder and TNT contain nitrogen. It takes a lot of energy to break nitrogen bonds, and a lot of energy is released when you do. To make a long story short, it's hard for living things like us, with our low-temperature biological processes, to pry nitrogen away from itself, so we use it only when we have to; in protein, for instance. As prevalent as nitrogen is, you and I—and our little mouse friend—are only 3 percent nitrogen."

It made perfect sense.

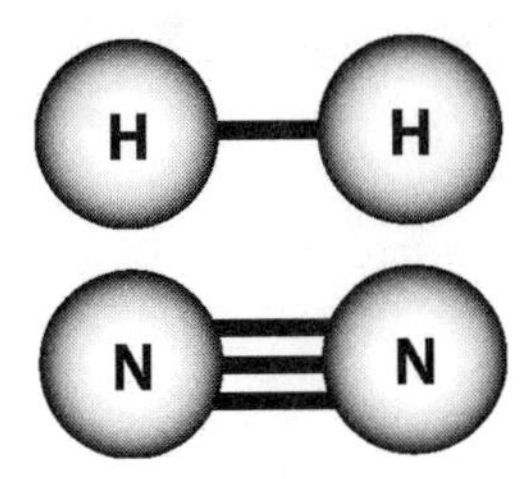

Weak vs. Strong Bonds

Hydrogen molecules share only 2 electrons and are easily broken. Nitrogen molecules, which share 6 electrons, are not easily broken and require a lot of energy to separate.

Though I felt silly asking it, there remained one last question. "I once saw an episode of Star Trek where there was a 'silicon-based life form' that tunneled through solid rock. The idea was that silicon was similar to carbon, and therefore it might be possible to build life around it. Considering how prevalent silicon is on Earth, I was just wondering..."

Zed beamed. He replied, "Oh, yes...I remember that one. Doc McCoy healed one of the creatures with some sort of silicon-based cement. Amusing show, but silicon-based life wouldn't work, I'm afraid, even though silicon does have many carbon-like properties."

"If it's so similar, why wouldn't it work?" I asked, truly curious.

Zed thought for a moment, then made a stick-and-ball figure appear in the air. It differed from the one he made for water in that all the balls were close to the same size and lined up in a row. "Grab it!" he said, "grab it, if you can! It's carbon dioxide."

It floated away, as I knew it would. Then he made the same stick-and-ball figure, with the exception that the middle ball was noticeably larger. "Okay. Try this. It's silicon dioxide."

I lunged and grabbed it, with no difficulty at all. I squeezed my fists together and felt very foolish as grains of white sand poured through my fingers.

"Too heavy, that silicon," Zed said. "It's carbon, or nothing."

I looked at Zed and smiled. He certainly had a way of getting his point across. Although this lesson was regrettably brief and rudimentary, I was beginning to develop a keener sense of how the world worked, and why. Because the chemistry of hydrogen is so thoroughly bound up with the chemistry of oxygen and carbon, you can hardly speak of one element without speaking of the other two, a fact that was born out again and again in the coming days.

Zed sez

With carbon, oxygen and hydrogen you can make all the sugars, fats and carbohydrates in the world!

technistoff - 4

All my college chemistry courses would have been immeasurably easier if, on the first day of day of class, the dour-faced instructor would have swaggered to the chalk board and written:

basics of molecule bonds

> Hydrogen likes to form 1 bond
> Oxygen likes to form 2 bonds
> Nitrogen likes to form 3 bonds
> Carbon likes to form 4 bonds

With those four simple rules one of the central tenets of molecular formation would have been laid bare, and anyone who ever wondered why water was H_2O instead of H_3O, and ammonia was NH_3 rather than NH_2, would need wonder no more. Suddenly the basic structure (bond angles, physical properties, etc. notwithstanding) of everything from methane (CH_4) to nitroglycerin ($C_3H_5O_9N_3$) would begin to make sense.

Basic Formation of Molecules

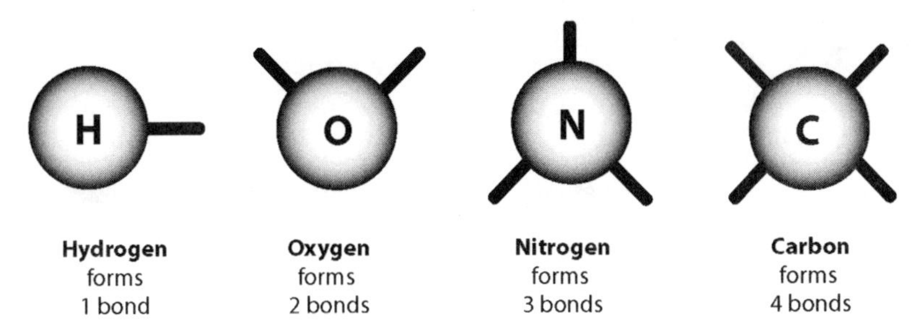

The fact that Zed was the first teacher to ever illuminate these basic rules is ironic, since it was not Zed's purpose to conduct a thorough chemistry course. He merely wished to explain, with his customary flair, the fundamental dynamics of the chemistry of life, and the pivotal importance of carbon in making life possible.

carbon implies life > life implies tissue > tissue implies food > food implies fuel

That this is relevant to the study of hydrogen power is underscored by the fact that carbon implies life, life implies tissue, tissue implies food, and food implies fuel. And fuel, be it sucrose ($C_{12}H_{22}O_{11}$) or propane (C_3H_8), is what brought us to this point in evolution where we can at last ponder the promises and complexities of hydrogen power.

While there were no groundbreaking revelations in this lesson, I should name a few sources I used to verify, and clarify, Zed's notions of chemistry.

After Zed explained the rules of chemical bonding, I found it written in very much this same form in the *Scientific American Library* book, *Molecules*, by P. W. Atkins. This deceptively simple and useful book also contains an excellent explanation of the hydrogen bond, and why it is so necessary for any number of chemical phenomena, up to and including the development of life.

Exploring Chemical Elements and their Compounds, by David L. Heiserman, is a great source for anyone wishing to learn about the individual elements, and their varied roles in the grand scheme of life, the universe, and everything.

While Zed's ball and rod representations of molecules was a useful tool, molecular bonds, when shown in three dimensions, are more generally displayed simply as balls stuck to one another, their sides flattened where they meet. In this way, water more closely resembles Mickey Mouse, than the bug's head I made with the balls and rods.

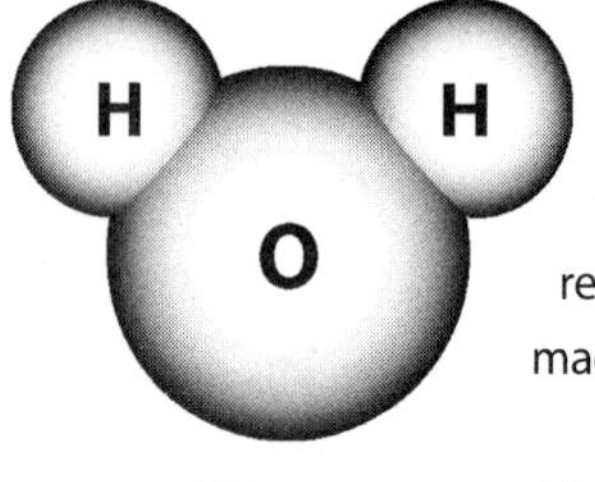

Water
H_2O

Graphical correctness aside, I can't imagine how this lesson would have turned out if Zed had simply presented me with a handful of flattened balls.

Zed sez

The positively charged end of hydrogen attracts the electrons from oxygen atoms in nearby molecules like ambulances attract lawyers. The end result is that every water molecule is drawn toward every other water molecule.

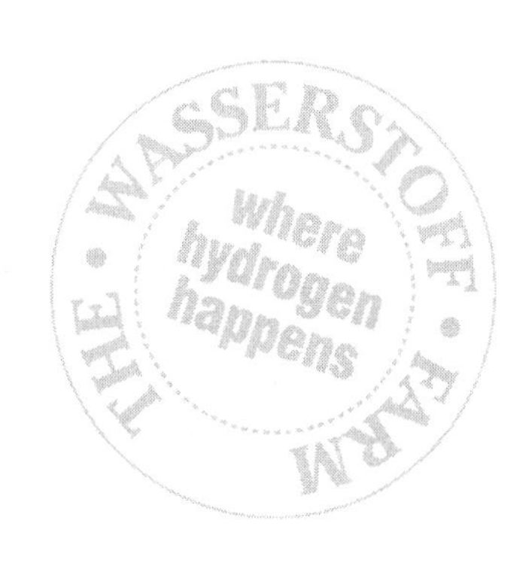

chapter 5

The Dangers of Riding Your Carbon Cycle Through a Greenhouse

It was another beautiful day at the Wasserstoff Farm. Unlike yesterday, Zed wasn't standing at the edge of my bed at the crack of dawn, and I hadn't had any unsettling dreams about weirdo beings from the planet Wasserstoff. Two things to be thankful for, and I hadn't even showered yet. The day was showing some promise.

After a long, hot, soothing shower, I was surprised to find my clothes—an old flannel shirt and a faded pair of Levis—had been laundered and pressed, and breakfast had been delivered to my cottage: orange juice and coffee; pineapple, bananas, and kiwi fruit; toast with grape jelly; something resembling granola, and...bacon? It was too good to be true. I grabbed a piece and hungrily shoved it in my mouth. It sort of tasted like bacon. Bacon and deep-fried tofu, maybe. Or just the tofu with a liberal dose of flavoring? I didn't care; it was the closest thing to meat I'd seen in two days and I wolfed it down before moving on to the undisputedly vegetarian fare.

Sated at last, I stepped out into the day. It was warm and bright—though a bit breezy—with the sun shining through a thin layer of haze; just like yesterday. Besides being unseasonably warm, I thought it unusual to have two hazy days in a row. In Colorado, it was usually either thick clouds or bright sunlight, with very little in between.

Whatever. I was getting used to the notion that what the Wasserstoff Farm was, and what it appeared to be, were not one and the same. Zed liked to keep a keen edge on his mysteries.

Where was Zed, anyway?

As if telepathically summoned, he quickly appeared from around the corner of a greenhouse. He was, oddly enough, riding a unicycle and juggling little red and white balls. "Dang! There goes another, one!" he exclaimed.

He must've been dropping his balls, I thought—what he was doing couldn't be easy—but as he rode closer I saw it was quite the opposite: for every five balls he threw in the air, only three or four came back down. The rest hung in the air a few feet over his head, before beginning to descend at a snail's pace. As balls were lost, new ones materialized in his hands. After a couple minutes, he was riding around in a cloud of slowly moving little two-tone balls.

As hard as I tried, I couldn't divine what the lesson might be in this spellbinding display of dexterity. Then I noticed that the balls weren't exactly spherical; they were agglomerations of two white balls smushed into the opposing ends of the red balls. Then I knew: it was carbon dioxide.

"Let me guess," I hollered at him, "you're demonstrating that human activities are putting more CO_2 into the air than nature is working back out of the system. This, in turn, is causing the atmosphere to heat up, inducing a planet-wide rise in temperature, right?"

He raked me with a curious glance and immediately stopped juggling his odd little balls. As he slid off his unicycle, all the balls vanished. Walking toward me, he said, "You drew that profound conclusion, just by watching me juggle a few balls?"

"Yep," I answered, proudly.

"Don't you suppose that just might be a little simplistic?"

He was dressed the same as yesterday—jeans and that strange lab coat of his—but he was wearing a different T-shirt. This one said: *Hydrogen Happens at the Wasserstoff Farm.*

Feeling a small but discernable bruise in my confidence, I replied, "No. Global temperatures are on the rise, and so are atmospheric levels of CO_2. It's pretty obvious, isn't it?"

He scratched his head, and answered, "Well, you'd think so. But the fossil record sometimes tells a different, even contradictory, story. It points to periods in the past when temperatures fell as CO_2 levels were rising. And vice-versa. Nature is funny that way; she enjoys outwitting us stupid humans."

I was confused. I asked, "How is that possible, Zed?"

He produced one of his lopsided white and red balls from out of the air, and gave it a toss. As if shot out a rifle, it whooshed along a trajectory-less course for a few dozen yards, before stopping in midair. The ball was attached to his hand by a taut, black string with thin, white gradations on it. He said, "All right, let's see what we're dealing with, just to get a sense of proportion. This string is 100 feet long. It represents the sum of all the various gases in the Earth's atmosphere. Why don't you show me how much of it is CO_2?"

I had to be careful. I knew that nitrogen would take up a little better than 78 feet, and oxygen would gobble up almost 21. That didn't leave much more than a foot to work with. I pointed to a spot about 3 inches from his hand, and said, "There."

Zed shook his head. "Are you trying to cause trouble, lad? We might really be in a stew if there was *that* much."

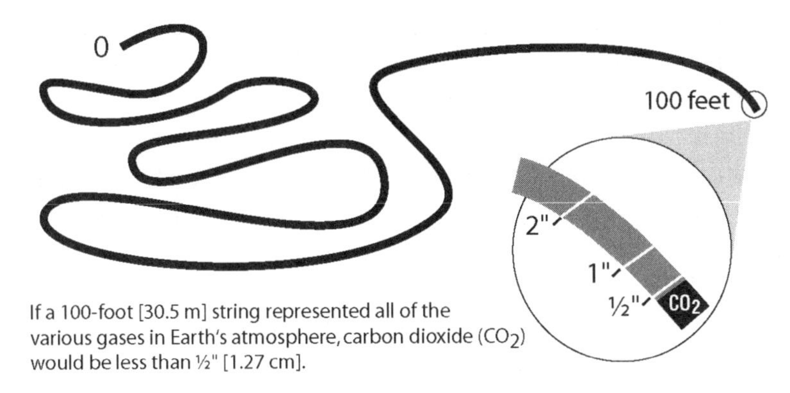

If a 100-foot [30.5 m] string represented all of the various gases in Earth's atmosphere, carbon dioxide (CO_2) would be less than ½" [1.27 cm].

Percentage of CO2 in our Atmosphere

I pulled my hand away as if the string were electrified. Gingerly, I placed my finger back down on the one inch mark. "How about there?"

"Ummm. Better, but you're still being reactionary. The fact is, atmospheric CO_2 right now is a little shy of the ½ inch line—0.44 inches in 100 feet, if you're a stickler for precision. That's up from 0.33 inches, just before the Industrial Revolution. Over 30 percent in the last 250 years, and most of that in the last 50 years."

"That's significant," I concluded.

"Indeed it is," he agreed, "and I would say we just might be headed for a scorcher of a future. But who knows? We have to accept the fact that the Earth is far more subtle and complex than we give the old girl credit for."

I was getting frustrated. I asked, "So just what *are* you saying, Zed? You're the caretaker of this giant complex, ostensibly existing for no other reason than to teach me about all the ways hydrogen can be used to replace fossil fuels, and yet you stand here and tell me you don't absolutely believe increased levels of CO_2 are going to turn the Earth into a hothouse? It makes no sense!"

He glowered at me, and barked, "*Caretaker?*"

Oops. "Sorry, Zed...I just..."

"Never mind. You have rudeness in your blood; I can accept that, if you can. As far as 'believing' goes: this isn't a religion, you know. Belief has nothing to do with anything. In all probability the current rise in CO_2 is largely to blame for the measurable rise in global temperature. As close as anyone can tell, the two phenomena are dancing in lockstep, and have been for at least the last 400,000 years. I'll accept the data as conclusive. But you need to understand that Earth's time frames are far more expansive than ours. What's going to happen down the road if we don't stop our nasty habits of dumping things into the atmosphere that don't belong there? Who knows? It might bring on the next ice age."

Zed sez

What's going to happen down the road if we don't stop our nasty habits of dumping things into the atmosphere that don't belong there? Who knows? It might bring on the next ice age.

"*Ice Age?*" I shrieked, "You've *got* to be kidding!"

His eyes flashed an arctic blue, "I never kid about ice ages, lad."

"But Zed..."

"All we can say for certain is that our predilection for burning things that would be better off left in the ground is going to do *something*. Are we clear?"

'We' weren't at all clear, but I mumbled, "Sure."

"Good. Now step into my office. This will be your final propaedeutic before we begin the study of hydrogen energy in earnest."

"My final *what?*"

"Oh, for heaven's sake, just follow me!"

I did. He led me into the large greenhouse near where he'd been riding his unicycle earlier. I noticed that the unicycle, which he'd left lying on the ground, was now nowhere to be seen.

Inside, the greenhouse seemed normal enough. It was sweltering hot, and smelled like a swamp. Beads of sweat began to form on my forehead and run down my cheeks. Looking around, I noticed there were no tables or pots. Plants of every variety grew strong and healthy, right out of the ground. The plants on one side I could identify—sunflowers, raspberry bushes, mountain grasses, juniper trees—but those on the other I could not. They seemed like plants I'd seen in drawings in natural history books—extinct varieties of giant ferns and strange-looking, conifer-like trees with long, shaggy needles. I seemed to remember they were called Lycopods. I almost expected to see dinosaurs lurking behind them. The thought made me uneasy. The ancient plants all grew out of bogs and we had to walk on a raised platform to keep from getting our feet wet. It was a thoroughly anachronistic garden.

Sensing my discomfort at seeing the long-extinct plants, Zed said, "Don't worry; this isn't Jurassic Park, you know. Way too early in time for that. I just wanted to add a little flavor to titillate your senses."

"Uh huh," I said, unable to hide my doubts.

"Really," he assured me.

"Glad to hear it," I told him.

"And to make a point," he added.

"What point would that be, Zed," I asked with no small degree of hesitancy.

"Watch this," Zed instructed. He snapped his fingers and instantly my breath—and Zed's—became a cloud of the same little white and red balls he'd been playing with earlier. They were much smaller, of course, impossible to see individually, and, while some hung in the air, most were quickly gobbled up by the plants. Suddenly, the plants began to spew out little blue double-spheres. In a moment the room was awash in colorful floating molecules. The blue ones were by far the most prevalent, but my eyes were drawn to the white and red ones with great interest. It seemed that however many of those we breathed out, the plants took them in. No more, no less. "A perfect stasis, or so it seems," he said, smiling broadly. "Plants take in CO_2 and along with water, create sugar and oxygen. We eat the plants, using their sugars, fats and proteins, and return CO_2 back to the atmosphere. Of course, even on a scale as small as this, it's nearly impossible to maintain the right balance. Something always gets out of whack. What looks perfect now could be a disaster inside a month. So imagine what it's like to do this on a planetary scale!"

"But why the two types of plants?" I asked. "What's the point in that?"

He smiled, and said, "Simply this: that both sets of plants come from geological periods when global temperatures and atmospheric CO_2 levels were pretty close to the same. This is interesting, because the plants on the right, as you may have guessed, are from the Carboniferous Period, some 300 million years ago. It was a time of luxuriant plant growth, when animals were few, ice covered much of the land, and ancient forests flourished

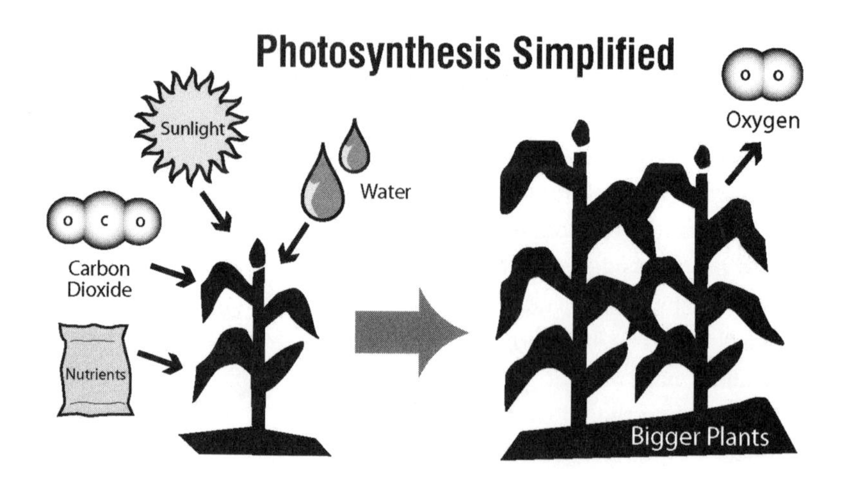

Zed sez

All we can say for certain is that our predilection for burning things that would be better off left in the ground is going to do something. Are we clear?

in the bogs and swamps of the tropics. It was also a time when much of the world's coal and crude oil was formed. From that time—all through the era of the dinosaurs—until relatively recently, global temps and CO_2 levels had their quirky ups and downs, but were, without exception, markedly higher."

I didn't know what to say. Staring into my blank face, Zed said, "Anyway, this is just the vestibule. The real action is in the back."

In the back? I'd seen this place from the outside and there wasn't enough space for any room "in the back." There was a door, however, and Zed led me to it. "Watch your step," he said, before opening it. I thought surely it would lead back outside, but I was wrong. Really, really wrong.

We stepped out onto a narrow, steel platform. It seemed to go all the way around the spherical room (if, indeed, the cavernous space before me could be called a room), but it was so vast, and the lighting around the sides so dim, my eye was unable to follow the platform's slightly curving shape for more than a few dozen yards. Nor did I care where it went, for as I gazed into the void looming before me, I found myself at eye-level with...the Earth. Or at least the grandest reproduction of it ever made. It hung weightless in the air—no wires or supports of any kind. It must have been 300 yards or better in diameter, though it was impossible to get any sense of scale, since I was unable to discern how far away it might be. For all I could tell, it was a mile across and three miles distant.

The detail was breathtaking. The polar regions gleamed like sculpted ivory, and the oceans shone and shimmered with every hue of blue and green. Snow-covered mountains hovered above parched deserts and verdant green forests, while population centers appeared as dark, smog-enshrouded places that brought to mind foreboding scenes of Mordor. Above it all, giant weather patterns danced across the planet in graceful, ever-changing swirls.

"Wow!" is the only word I could manage.

Zed replied, "Well put, lad. Now, before we begin, could you please explain to me the concept of the 'Greenhouse Effect'?"

I thought for a moment, then answered, "Well, as I understand it, sunlight travels through glass—as in the roof of your greenhouse—as a relatively short visible wavelength of the light. This light is absorbed by things within the greenhouse and re-emitted as a longer, less energetic, infrared wavelength that is unable to get back out again. The heat is absorbed by the molecules of the air and the greenhouse heats up."

"Very good! Now, how does that apply to our planet's atmosphere?"

I answered, "I suppose it's a lot the same. Of course you have clouds that reflect light back into space..."

"About 25 percent of it," Zed interrupted.

"...and some of the light must be absorbed by the atmosphere..."

"Let's say another 25 percent."

"So that means about half the sun's radiation actually makes it to the Earth's surface?"

"More or less," he agreed.

"Most of which is absorbed by the Earth...?"

"Correct."

"While a lesser portion is reflected back?"

Zed took over from there. "Right. Part of this radiation—both the reflected and the absorbed—is released back into the atmosphere as heat, and is absorbed by greenhouse gases. These gases—most notably water vapor, but also CO_2, methane, and others—absorb the infrared, which is to say heat radiation, and emit it in all directions. Some is lost into space, some is radiated toward Earth to help warm the planet. It's what makes Earth habitable. Without greenhouse gases the average surface temperature would be very close to zero. Fahrenheit, that is. A rather Martian scenario, I'd say."

I had a feeling Zed was about to make a demonstration, so I prompted him, saying, "But I take it the greenhouse effect isn't the only thing driving the Earth's climate?"

"And you would be right." He turned toward the spinning globe before us, and cracked his knuckles as a orchestra director might, just before a concert. "You see, for most of the Earth's history, the concentrations of CO_2 have been much higher than they are today, and temperatures have been much hotter. Global temperature and CO_2 levels have often not risen and fallen in the same direction at the same time, however, leaving the impression that there isn't always much correlation between them. That's where other factors come in..."

His fingers went into a dance, and I watched, mesmerized, as the planet began to change before my eyes. Ice sheets spread over the northern and southern hemispheres, receded, and spread again. The continents pulled apart from one another, drifted in a seemingly random fashion, and smashed together again, forming huge mountain ranges as they did so. Forests grew from deserts, only to be reclaimed again by the encroaching sands. Ocean currents—visible as intense blue rivers within the seas—reached from the equator to the high and low latitudes, only to be cut off by moving land masses, which were shortly covered again in ice. Through it all, the intensity of the light hitting the Earth waxed and waned

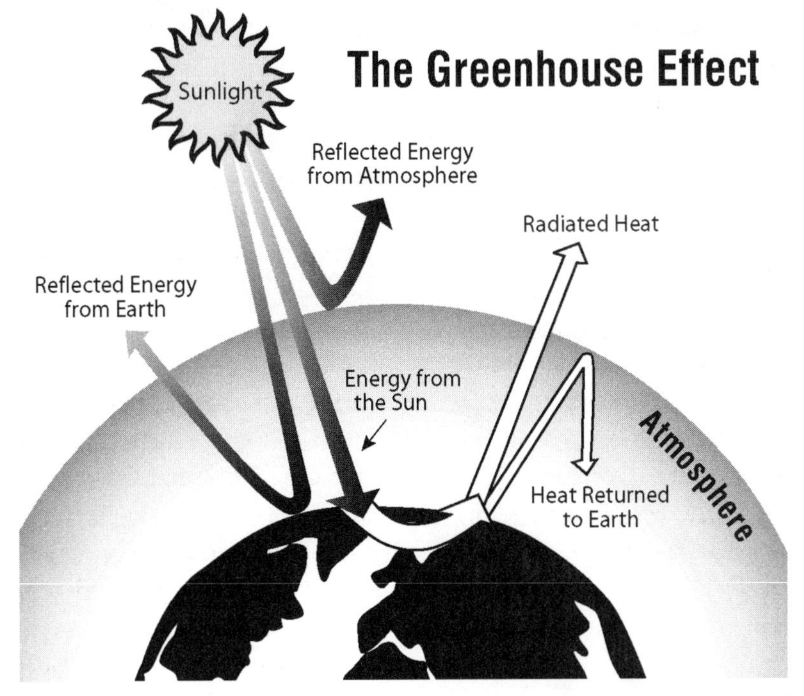

The Earth is covered by a blanket of gases (our atmosphere) which reflects about 25% of the sun's energy. The remaining energy that reaches Earth is either reflected or converted into heat, which keeps us from freezing. Human activities, such as burning fossil fuels and land clearing, are dumping more heat-absorbing greenhouse gases (such as carbon dioxide and methane) into our atmosphere, which in turn, trap more of the heat.

almost imperceptibly, and the Earth repeatedly wobbled a few degrees back and forth on its axis.

"This is Earth's past, and its future. Because of gravitational influences from Jupiter, Saturn and Venus, the Earth wobbles back and forth on its axis every several thousand years, changing the distribution of solar radiation upon the Earth's surface. These same gravitational forces periodically change the shape of Earth's orbit about the sun, making it vary from a near-circle to a more elongated ellipse. To further complicate things, the intensity of the sun waxes and wanes in much shorter cycles. And over much longer time spans, the continents are pulled apart by tectonic activity in the Earth's crust, and slammed back together in vastly altered arrangements.

"This, in turn, redirects ocean currents and the global distribution of heat. It's nothing we can prevent, even if we wanted to. However..." He pulled his hands close together and balled them into fists. The continents rearranged themselves into a recognizable configuration and the forests and deserts and ice caps returned to normal—at least as far as I could tell. "...It hardly gives us *carte blanche* to do as we please. Quite the contrary. While it may be true that, compared to the greater portion of Earth's history, the atmosphere currently holds lower concentrations of carbon dioxide, it's something every organism on the planet is quite comfortable with. And though the 30 percent rise in CO_2 over the past quarter-millennium may seem gradual to us, it's a blink of an eye, geologically speaking.

"In the last 400,000 years, atmospheric CO_2 concentrations have risen and fallen several times—from a high of 300 ppm some 325,000 years ago, to a low of around 190 ppm at several other points during the period. But now it's at 370 ppm and not likely to drop anytime soon. Before we humans came onto the scene, the only events that could pro-

Zed sez

Without greenhouse gases the average surface temperature would be very close to zero. Fahrenheit, that is. A rather Martian scenario, I'd say.

duce such quickly elevated levels of CO_2 were catastrophic ones, such as volcanoes..." I watched as dozens of bright red pimples appeared on the land and grew out of the sea, belching huge plumes of dense, black smoke that began to swirl around the globe in the direction of the prevailing winds. "...Or comets, meteors and asteroids."

From somewhere behind my head, a large, pitted rock whistled into the atmosphere and burst into flames. When it hit the ocean near the tip of the Yucatan peninsula, a giant tsunami appeared out of a fireball of steam and ash and swept out in all directions, engulfing huge tracts of land on four continents. A roaring and devastating pressure wave followed. It spread all around the globe, causing most of the world's forests to burst into flames. Soon the entire globe was shrouded in steam and smoke and dust, and nothing familiar remained for the eye to see.

"The point is," he went on, with what I took to be a disturbing nonchalance for a man who'd just destroyed a planet, "nature always finds a way back. But never as she was before."

Still in shock, I watched as the smoke began to clear. Momentarily, the ice caps disappeared and the continents shrunk noticeably as the sea level steadily rose. The old pattern of forests and deserts was also different. I wondered what creatures and plants might inhabit this strange, new world.

Zed shook me from my dream state. "The bottom line? We're digging up carbon that's been buried deep in the ground for a quarter of a billion years and spewing it back into the atmosphere as if forced to do it by a pantheon of malevolent gods. It *will* change things, that much is for certain. How much, and in what way, who knows? Until the last 250 years, our planet was in a most exquisite balance; it really was miraculous. But now...?"

I hated to interrupt his thoughts, but something was nagging me. I asked, "Uh, Zed? Earlier you mentioned this little matter of an ice age? Do you suppose we might talk about that?"

"Why not?" he asked, with narrowed eyes. "Always time to talk about an ice age. It's one of my favorite subjects. The next one, you should know, has to do with the Gulf Stream. Ever hear of it?"

"Sure," I said proudly, "it's the big stream of warm water that flows from the tip of Florida into the North Atlantic. It's what keeps Europe's climate so moderate."

"Good," he said, approvingly. "There may be hope for you yet. It works like this: the warm tropical waters of the Gulf Stream begin to evaporate as they flow north. This makes them saltier, and thus heavier. Add to this the cooling effect of winds coming off the North American continent, and you have cold, very salty water. It sinks to the ocean floor and flows south as a deep ocean current, drawing more warm water up from the tropics. It's a great system for getting heat from a warm place to a cold one."

"Sounds like it," I agreed. From the corner of my eye I saw that the Earth had returned to normal; at least the current version of normal. The Gulf Stream was highlighted as a vibrantly glowing blue stream of water flowing from the Florida coast to just south of Greenland. At this point it suddenly plunged to the depths and headed south, down the middle of the Atlantic, clear to Antarctica, where it fanned out into the Indian and Pacific Oceans, before heading north again.

"So tell me," he continued, "what would happen to Europe and the upper regions of North America if the Gulf Stream were to suddenly shut down?"

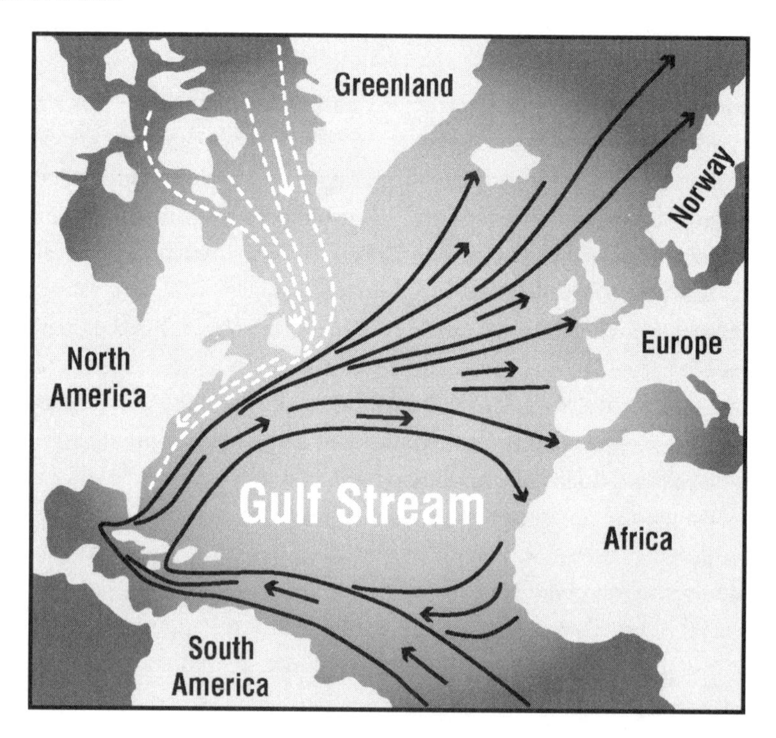

"They'd both get pretty cold, I'd say. But," I objected, "how can that happen when the planet is getting warmer? The Gulf Stream should just get bigger and hotter, right?"

"Indeed it should. And if it weren't for the nagging matter of a few hundred thousand cubic miles of ice to the north, it just might. But what if all that ice started to melt and flow into the North Atlantic? Besides raising sea levels—which seems to be the biggest concern, these days—it would change the salinity of the northern seas. The water would become lighter and more buoyant. It wouldn't sink—it would just sit there and freeze. A marvelously-crafted, planet-sized heat distribution system would cease to exist. Get it?"

Oh, I got it, alright. I watched with equal measures of curiosity and trepidation as the coastlines of the world receded ever so slightly as the arctic ice sheet and the Greenland glaciers began to recede. Then, in an instant, the Gulf Stream disappeared. The serpentine loop of water—many times bigger than all of Earth's rivers, combined—had stopped flowing. I was horrified.

But things soon got worse. Quickly the northern ocean froze over and the ice sheets that only moments before were receding, began to expand. In an instant—which may have been a hundred, or a thousand, or a hundred thousand years—the ice spread from the polar extremes, all across northern Europe and Asia, crept down the North American continent toward the Great Lakes, and continued south, gobbling forests and centers of civilization with equal relish.

I was dumbfounded.

I jumped when Zed tapped me on the shoulder. I turned to see him staring me down with squinted eyes, his jaw thrust into a masculine set. In a voice that was soft, low, and menacing, he said, "So the question is: are ya' feelin' lucky, punk?"

"Not lately," I said, marveling at his dead-on Dirty Harry imitation.

His whole countenance changed as he put his arm around my shoulder in a fatherly way, and said, "Me neither. Let's go see what we can do to improve the odds, shall we?"

technistoff - 5

Zed often raises more questions than he answers, as any competent teacher should. But a great teacher, like Zed, will also instill in the pupil the energy and curiosity to make him run out of the classroom and start turning over stones to see what lay beneath them. Keep the flow of knowledge moving, as it were.

Nature is flow. And ebb. And sometimes, hidden traps. That was the point of Zed's lesson. Or, at least I think it was. He will, after all, forever be an enigma...

Now for what I was able to find in my quest to verify his claims:

CO_2 & global warming

A wonderful graph of the rise in atmospheric concentrations of CO_2 since the beginning of the industrial age can be found on the web at *www.grida.no/climate/vital/07.htm*, though you should be warned it's in PPM, instead of hundredths of inches on a 100-foot string anchored to Zed's palm.

For a concise and colorful graph showing the rise and fall of global temps and CO_2 concentrations for the last 600 million years, visit *www.geocraft.com/WVFossils/Carboniferous_climate.html*. It's telling, to say the least, but just *what* it is telling us is hard to interpret. The grand, and ultimately muddy, lesson is that

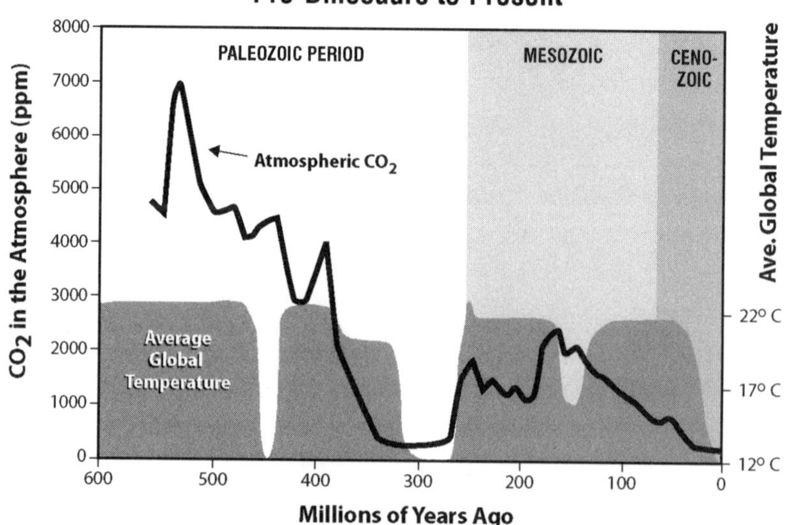

Earth always figures a way out her difficulties, though we often don't know how. Nor does she do it in ways most of us would find comforting, considering we have built a colossus of a civilization based on a multitude of narrowly defined climatic parameters.

On a more contemporary note, a thoughtful, even-handed article on global warming can be found in the March 2004 issue of *Scientific American*. Written by James Hansen, director of the NASA Goddard Institute of Space Studies, "Defusing the Global Warming Time Bomb" explains why we know what we know, and what we might expect the future to bring. He begins by comparing the extra amount of heat caused by the current high levels of greenhouse gases to the amount of heat that would be given off by a pair of 1-watt Christmas-tree bulbs, burning day and night over every square meter of the Earth's surface. Over time, Hansen convincingly explains, that tiny extra amount of energy could lead to drastic consequences.

other greenhouse contributors

While carbon dioxide is the single biggest contributor to the global warming engine, it is not the only one. Methane, chlorofluorocarbons, nitrous oxide, and ozone contribute to the global warming equation, as do minute particles of black carbon (soot, as from diesel exhaust)—which change the reflectivity of clouds and even the polar ice caps, causing them to absorb more energy and thus warm at a faster rate than the rest of the planet.

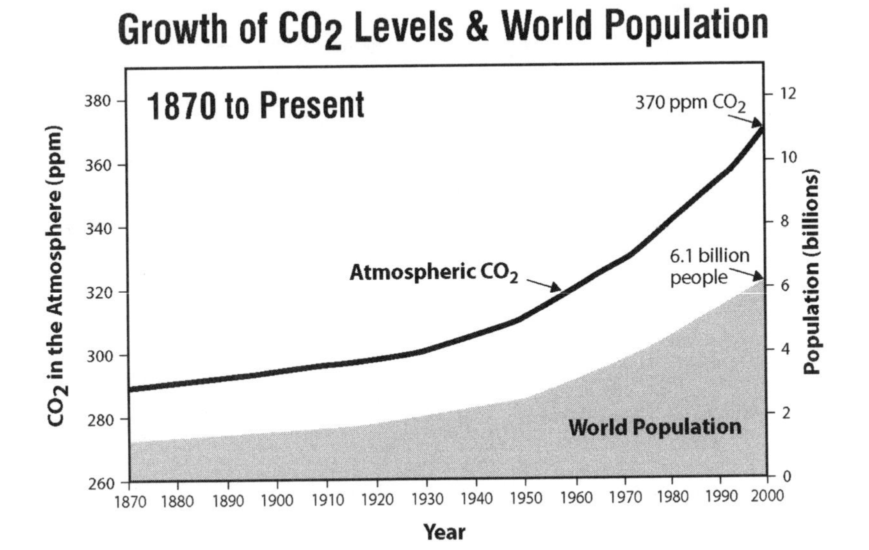

Ironically—and disturbingly—global warming is actually held somewhat in check by other human-caused "forcings." Among them, reflective aerosols (sulfates, from the burning of fossil fuels, for example) which reflect a portion of solar radiation back into space help cool the planet, as does the clearing of forests in northern latitudes, since snow cover is more likely to reflect sunlight from snow-covered crop stubble than from land covered with tall trees. And there is a final irony: the winds created by the ozone holes over Antarctica and Greenland, Hansen concludes, have protected these ice sheets from rapid melting. As the ozone holes heal up—due to the phasing out of chlorofluorocarbons—we can then expect the melting to accelerate.

To read an expanded version of Hansen's article, visit the Goddard Institute for Space Studies at: *http://pubs.giss.nasa.gov/docs/2003/2003 _Hansen.pdf.*

the gulf stream

Long before I ever met Zed I had the good fortune to run across an article in the December 1996 issue of *Discover Magazine*, concerning the Gulf Stream. Written by Robert Kunzig, "In Deep Water" details in refreshingly lucid prose just what the Gulf Stream is, how close we are to shutting it down, and what we can expect if that happens. Had I remembered it sooner, I might have looked a little less green-behind-the-ears in Zed's eyes. For a more recent (though hardly changed) account, read *The Ice Age Cometh,* by Thom Hartmann, at *www.alternet.org/story.html?StoryID=17711*.

It's a must-read for woolly mammoth enthusiasts.

The Clean Urban Transport for Europe Project (CUTE) is doing its part to reduce the global greenhouse effect by using fuel cell hydrogen buses throughout Europe. This Amsterdam bus is just one of many that demonstrates an emission-free, low-noise transport system. For more information, visit *www.fuel-cell-bus-club.com. Photo courtesy of Shell Hydrogen; www.shell.com.*

chapter 6

Deeper into the Wasserstoff Farm

After what Zed had just put me through, I needed a break. Intentionally or otherwise, the snowy-haired old Druid had dumped more information, conjecture, and visual imagery into my mind than my mind was designed to hold. My head felt much like my stomach always did after Thanksgiving dinner at my grandma's house. Rather than ask me if I'd care for second helpings of every dish in her 20-course meal, the sweet and smiling little woman would simply pile food on my plate. Then she'd grace me with a well-meaning stare until it was all gone. I customarily left the table looking and feeling like I'd swallowed one of her prize pumpkins.

Grandma and Zed had a lot in common.

So, fearing that my head might split open unless my brain had time to digest the last lesson, I asked Zed if I could take a ride around the Wasserstoff Farm on my content, but neglected, horse. He seemed to hesitate, but finally said, with furrowed brow, "Just stay near the perimeter, if you would. The interior might prove to be...dangerous."

Dangerous? What could be dangerous about a benign enclosure run by an eccentric old sorcerer, where the grasses and trees were always green and the flowers bloomed in late fall? Even my skittish horse, who habitually saw goblins hiding under every rock, thought it was paradise.

Just the same, I took Zed's advice—for awhile, at least.

Against all expectations, Mike seemed happy to go exploring, so I let him have his head as we took a leisurely walk just inside the wall. I appreciated Mike's newly-acquired demeanor, because I was in no mind to wrestle with a fractious horse. I just wanted to think.

Zed's take on the global warming matter was...interesting. And, I had to admit, more than a little disturbing. In my short experience with the issue, I'd been led to believe there were but two sides to the global warming question: those who thought it was a chiseled-in-stone incontrovertible fact that the globe would continue to get progressively hotter, ushering in a plague of climatic changes that would ultimately doom humanity; and those who had convinced themselves it was all a lie—a tree-hugger conspiracy, vindictively fabricated to lay a guilt-trip on all of civilization. The former relied on sophisticated computer simulations of the changes in climate we can expect to see over the next century (given the current state of knowledge), while the latter cavalierly pointed out that, for much of the Earth's history, there has been but a tenuous correlation between greenhouse gas concentrations and global temperatures. So why should there be any now?

The whole matter had, regrettably, decayed into a venomous political spat. On the one hand, experience has shown that computer models can't even predict next week's weather. On the other hand, for those who don't acknowledge a correlation between greenhouse gas concentrations and global temperatures, it's a fact that the two have risen and fallen in lock step for the past half-million years. For the two opposing sides, it was war: do whatever it takes to silence—and sully—your opponent.

I found the whole thing disconcerting. The planet didn't need that kind of divisiveness; in the end, Earth would do what she would do. Zed understood this. In that graphically memorable way of his, he simply pointed out that we are significantly changing several of Earth's key climatic parameters, and if we like things the way they are, it would behoove us to quit. Yes, it might turn out exactly as the climate-modelers say it will. On the other hand, it might not; it might end up twice as bad as anyone thinks. Or entirely the opposite of what anyone thinks.

There is nothing crude or simplistic about planet Earth. She dances and sings to a million different harmonies. From time to time some harmonics cancel others; other times they dance together. There are more triggers to Earth's climate than anyone can imagine, and it's pure folly to believe we can possibly know how heavily we can tread on each and every one before setting off a chain reaction, the results of which might make us all look back with nostalgia at the notion of global warming.

The foul-tempered house cat—with its claws and teeth of known quantity and dimension—may bite and scratch and inflict a lot of pain and misery, but it's nothing compared to what that wounded, vengeful tiger hiding in the bushes can do...

I ended my ruminations abruptly, for it suddenly occurred to me that Mike and I were getting nowhere. As I suspected from the beginning, there were no discernable breaches in the circular wall, at least not in the short distance we'd covered, which was at best but a small fraction of the Wasserstoff Farm's circumference. The place was huge; at least 5 miles in diameter, I reckoned, and maybe more. A lot more.

What sort of facility could possibly require so much space? All the buildings I had seen were within a relatively small area near the gate; the rest was simply rocks and trees, bushes and grass. And yet it was all surrounded by a towering, impenetrable wall that was curiously crenellated at the top. The mystery was more than I could bear. Against Zed's admonitions, I turned Mike and headed into the interior.

For the first half-mile or so, everything was the same: mountain flora with the occasional rock outcropping jutting up from a naggingly flat plain. It was the penultimate Ponderosa pine forest—large widely-spaced trees, the spaces dotted with a healthy smattering of grass and small bushes. The kind of setting you'd always see in those old western TV shows. The problem was, I knew these hills and there weren't five flat acres—at least not attached to each other—anywhere within 30 miles. So how had Zed found several square *miles* of flat ground, and then built an enclosure around it that would make the Great Wall of China look like a picket fence?

And then there was the wind. The deeper toward the interior we traveled, the stronger it became. The change was subtle but noticeable because—and this was yet another oddity—there were no gusts or swirls to speak of; it forever blew from the outer wall toward the center, yet, as I said, increasingly more forceful as we moved "inland."

The first change of scenery came about a mile from the wall. Abruptly the familiar mountain terrain gave way to vast orchards of fruit trees. Cherry, peach, and apple trees all grew well in Colorado I knew, but certainly not mango and papaya, as I was seeing now. And all the trees were laden with fruit. I rode under a peach tree and plucked a juicy-looking specimen. Did I expect it to be an apparition, or to taste like cardboard? I don't know; but I was pleasantly surprised to find it every bit as substantial and delicious as it looked. I then tried many of the other fruits, with the same result. This stuff was real!

I had the feeling I shouldn't be here, that Zed would be most displeased if he knew where I was, but I was in no mood to stop now. I rode through the orchard, further from the perimeter, sampling fruit along the way. Mike enjoyed the soft grasses that grew between the trees, and even the occasional apple. It would've been like paradise, I mused, except for the fact that the temperature had grown almost uncomfortably warm and the wind was becoming an annoyance.

It was at the interior edge of the circular orchard that Mike and I first heard the sound. The fruit trees gave way to vast fields of grain and other low-growing crops. I could pick out wheat and oats, potatoes and soybeans. I thought I could see corn in the distance, and perhaps even sunflowers, but Mike was so agitated by the high, whirring sound coming from beyond the fields that I felt it prudent not to push him.

That's when I noticed something very peculiar about the wind.

Perhaps Mike noticed it too, because he turned quickly and headed toward home. For once, my horse and I were of the same mind.

chapter 7

Sizing Up the Beast (with apologies to Rhode Island)

When I returned from my ride I found Zed sitting on a rock, wearing a fresh new T-shirt. It read: *Qualified Quad Quasher At Work.* He was lighting kitchen matches, one by one. From the looks of the pile beside him, he'd been at it for quite a while.

After unsaddling Mike, I slowly made my way over to where he was sitting. "Nine hundred sixty-nine, nine hundred seventy...oh, good grief! This is pointless!" He looked up and saw me. "Well, it's about time!" he said sharply. "I've been waiting for you to come and rescue me from this dreadful exercise!"

"Zed, what in the world are you doing?" I asked.

"Doing? What am I doing? Why, I'm making a point, of course."

"Of course. But what is the point you're making, if you don't mind my asking?" He seemed distracted, which probably meant he had a lot on his mind.

"Oh, that," he said. "Well, I was seeing how long it would take to equal the total

annual United States energy consumption."

"By burning kitchen matches?" I asked, incredulously.

"Well, yes, but I just don't seem to have the time." He was pulling my leg; had to be. "Of course, if we did it symbolically, then it might work," he concluded.

"Symbolically?" I chimed.

"Yes, you know, when something represents something else that it isn't. Otherwise this is just a fool's errand."

Thoroughly puzzled, I replied, "You've lost me, Zed."

Under his breath he mumbled, "That's no great feat," then looked up and smiled, saying, "all right, here's the story—the total U.S. energy consumption for the year 2002 was over 97 quadrillion Btu. Since the energy put off by a kitchen match is about one Btu, I thought it would make a good example. Now, there's 970 matches in that pile. If each match symbolically represents 100 trillion matches, then that's about right. Follow?"

"Ninety-seven quadrillion is a lot of matches, Zed."

"Don't I know it! At one match per second I would had to have started this exercise about 3 billion years ago, when life was first getting started on planet Earth."

"Probably didn't even have matches back then," I reasoned.

"Doesn't matter; wouldn't have burned anyway. No O_2."

I asked, "Just how big of a pile is 97,000,000,000,000,000 matches, anyway?"

"Pretty big," Zed answered.

"But *how* big?" I prodded.

"Well, if my calculations are correct, 97 quadrillion matches would cover the state of Rhode Island 56 feet deep in matches. All 1,214 square miles of it. That is, of course, if they're all neatly stacked. Dump 'em in there helter-skelter and the pile gets a lot bigger."

I could hardly believe what I was hearing. "That's incredible!" I exclaimed, "I had no idea this country used that much energy!"

Zed replied, "That's why they give it to you in 'quads'. It's nice and neat. Only 97 of 'em. Why, with only 97 you still get change back from your dollar. But start airdropping matches on Rhode Island and someone's bound to complain."

Changing the subject, I asked, "Why do they use Btus, anyway? With my wind and solar electric system I'm used to thinking in amp hours and kilowatt hours."

Zed answered, "A Btu is a unit of heat. It's the Brits' answer to the French calorie. While the calorie heats one gram of water one degree Celsius, a Btu, or British thermal unit, heats one pound of water one degree Fahrenheit. So, one Btu equals 252 calories. It also equals 0.252 kilocalories—or simply Calories—the kind you count when you eat sugar cookies. The irony to whole thing, of course, is that the Brits have since abandoned the English system of weights and measures, and adopted the metric system—a thoroughly French invention.

"But to answer your question, most of the energy used in this country is created by the production of heat. Coal and nuclear power plants create electricity by using heat to create steam, to turn turbine generators. Your pickup runs on gasoline which, when ignited with a spark, produces so much heat, so quickly, that the air inside the cylinders expands exponentially and forces the pistons down, thus turning the crankshaft. And, of course, you heat your house with propane or natural gas which produces...heat."

A thought occurred to me. I said, "So what you're saying is, if you use heat to produce energy you're also making pollution?"

"That's a bit simplistic," he cautioned, "but essentially true. Hydroelectric power, wind power, and photovoltaics all rely on the sun—since the sun drives the winds, and lifts the water to the high points above the dams—but produce no heat themselves. And no pollution. Geothermal, on the other hand, uses heat left over from the Earth's formation, but it, too, is non-polluting."

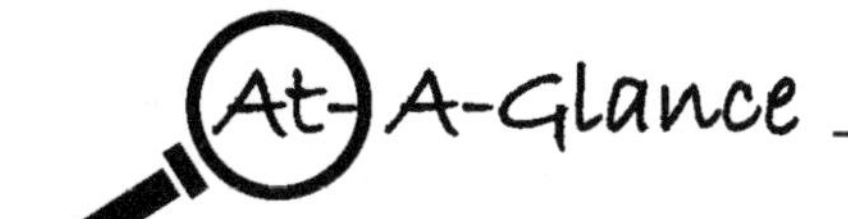

What is a Btu?

- ★ One Btu heats one pound of water one degree Fahrenheit
- ★ One Btu = 252 calories or .252 kilocalories (Calories)
- ★ ***Theoretically***: 3,410 Btu = one kilowatt hour (1 kWh) of electricity
- ★ ***Reality:*** It takes up to 11,300 Btu to produce 1 kWh of electricity *(due to losses and inefficiencies in production and transmission)*

"So where do kilowatt hours fit in?" I asked.

"Kilowatt hours are what the Btus produce," he explained. "Theoretically, it takes 3,410 Btu of heat to make one kilowatt hour of electricity. As an example, how many kilowatt hours of electricity does your solar and wind system produce?"

I had to think for a minute. "Well, on average, about 6 kWh per day, I'd say."

"Okay. That's a little over 20,000 Btu per day, or a whopping 7,466,000 per year."

"Wow!" It sounded impressive.

"That's equal to...let's see..." his fingers began to twitch, and he drew an outline in the air that quickly became a large block of stacked wooden matches, "...to about a 5-foot cube, excised from all those wooden matches in Rhode Island."

I looked at the cube, and exclaimed, "*That's ALL?!*" I was suddenly deflated.

Zed made the cube disappear with a wave of his hand. "But think about it," he said, "by using compact fluorescent light bulbs, efficient appliances, and turning off all loads after you're finished using them, you probably *saved* another 6 kWh's per day. If a million other homeowners did that, our 5-foot cube would become a kitchen-match monolith, 524 feet on a side."

"But still, Zed..."

"And, of course, that's not the whole story."

That caught my attention. "It isn't?"

"It goes back to the Btu/kilowatt hour thing. Remember I said 3,410 Btu would *theoretically* produce one kilowatt hour of electricity? Well, in practice it takes around 11,300 Btu, because about 70 percent of the heat and electricity is wasted in production and transmission. So now, suddenly, we can clear almost 150 of Rhode Island's football fields, provided, of course, it has that many."

"It doesn't seem like much," I lamented.

"Obviously, you've never cared for football. Doesn't surprise me, really. I could tell right away you're not a team player," Zed replied.

"That's *not* what I meant, Zed!"

"Of course it isn't, he-with-no-sense-of-humor. And you're right; it isn't much. It's only about 0.025 percent, or, if we can cut up Rhode Island into a million pieces, 250 parts per million. But that's not much lower than the CO_2 concentrations we're all concerned about, so it is significant in its own way."

He paused for a moment to study me, then said, "But enough with the matches. They were fun, but we should really bury poor Rhode Island under some *real* fuel."

"Such as?" I asked.

How about coal?"

"Why not?"

"Bituminous or anthracite?" he asked.

"Whichever one is used most," I decided.

"That would be bituminous. It's cheaper and more plentiful, and also less dense, so it's going to take more of it. About 4 feet, 9 inches, if we're dumping it there, or 2 feet, 11 inches, if we lay it in, in one solid seam."

"Dumping is easier," I concluded.

"True. Then it's 57 inches of coal on poor Rhode Island. Here's how it's going to be used: **18.92** inches for industrial use, to make the stuff that goes to the wholesalers and retailers for commercial use, which adds

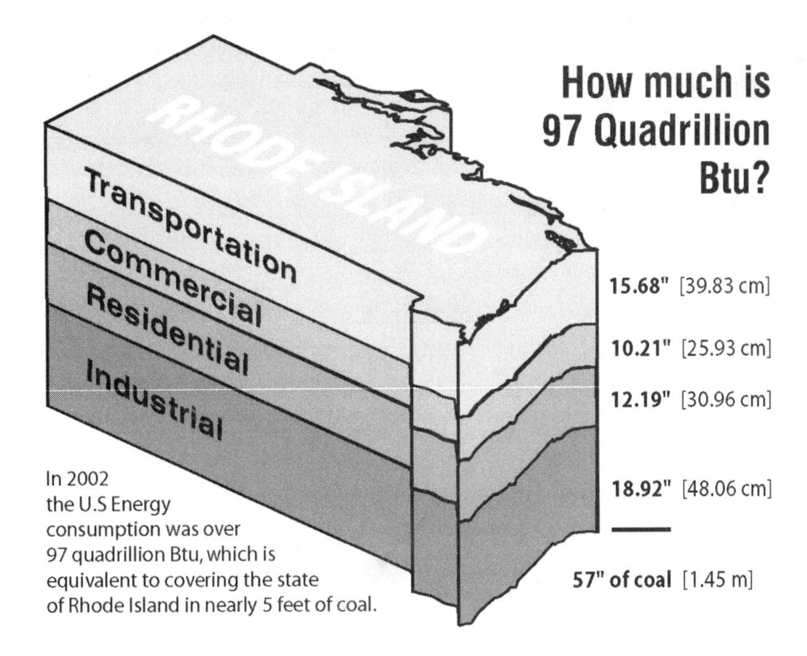

another **10.21** inches for the commercial sector. But, of course, you have to drive to the stores in the commercial sector, to get the stuff that was made in the industrial sector. And don't forget the trucks and trains and planes that move all this stuff from here to there. In total, we have **15.68** inches of loose bituminous coal on Rhode Island for transportation. That leaves residential. This is all the energy you burn at home, staying warm, happy and well-fed as you enjoy the stuff produced from the industrial sector, which you bought from the commercial sector, and brought to your home utilizing a means of conveyance from the transportation sector. This amounts to **12.19** inches for residential."

I shook my head. Where was he getting all these numbers? I had no idea, so I asked, "How about oil?"

"Medium grade, about 2 feet, 8 inches on 1,214 square miles," he answered, without seeming to give the matter any thought at all.

I wondered if he already knew these numbers, or if he was just that good at math, so I asked, "Okay; try this. Give me butter."

Without even raising his eyebrows, he said, "3 feet, 9 and a half inches, more or less."

That was past my belly button! A vast field of chest-deep butter, as far as the eye could see in every direction. What a concept! A 'Butter Economy' would surely be a shot in the arm for the dairy industry, I mused.

"Maple syrup," I said.

"That would be 7 feet, 2 and a half inches," he replied. "Sugar doesn't carry near the energy of fat." Probably a good thing, I thought; the world didn't have nearly enough maple trees.

I was impressed. Doubting now that I could stump him, I said, "How about hydrogen?"

Strangely, that seemed to stop him. Being as Zed was the proprietor of a place called the Wasserstoff Farm, I found it amusing. Then he made it clear I hadn't given him enough information. He asked, "At what pressure?"

"Huh?" I said. I hoped I sounded intelligence.

Apparently not. Again, he asked, "At what pressure? The stuff is a gas, you know! It doesn't just lie there in nice little cubes you can pick up and stack like bricks."

"Uh, no, I suppose not," I said. "Why don't you tell me what pressures it comes in."

"*What??* And you're supposed to be writing a book on this subject?"

"It's a ways off," I admitted.

"Well, *that's* good news."

Ignoring his rebuff, I said, "Okay, liquid hydrogen."

"Good. Now we're getting somewhere. Ninety-seven quadrillion Btu of liquid hydrogen would cover Rhode Island in 10 feet, 9 inches of the stuff. Essentially, it would be a shallow lake, 35 miles square, boiling at minus 423 degrees Fahrenheit."

"Wow!" I exclaimed, unable to contain my surprise at this enormous volume of liquid hydrogen. "And how much water would it take to provide that much hydrogen?" I asked, thinking it would be a deep lake, indeed.

"About 1.7 trillion gallons. It would make a lake out of Rhode Island 6 feet, 9 inches deep."

"What?! That's all? I mean, it sounds like a lot, but..."

"But you thought it would be deeper? Remember the hydrogen bond we spoke of earlier? If you recall, it gives water molecules polarity, and makes them mutually attractive. It's the reason water is so dense. Hydrogen molecules, on the other hand, don't give a whit about their own kind. No attraction, no density. A 5-gallon bucket of liquid hydrogen weighs about the same as a feather pillow."

He studied me, to see if he was getting through, then said, "I wouldn't be too terribly disappointed, if I were you. That much water—1.7 trillions gallons—is a little over 4 days, 9 hours worth of water flowing down the Mississippi River, dumping into the Gulf of Mexico at a rate of 600,000 gallons per second.

"But let's be fair. Even if we only wanted to use hydrogen as a fuel for the Transportation Sector—a logical assumption—it would still take the hydrogen in over 467 billion gallons of water, or the total flow of the Mississippi from 7:00 a.m. Monday, until noon on Tuesday."

Zed sez

Ninety-seven quadrillion Btu of liquid hydrogen would cover Rhode Island in 10 feet, 9 inches of the stuff. Essentially, it would be a shallow lake, 35 miles square, boiling at minus 423 degrees Fahrenheit.

My mind was swimming with huge, unwieldy numbers and seemingly impossible comparisons and analogies. And still I was having a hard time visualizing just how large a number 97 quadrillion really was. I told this to Zed, who attempted to surround the problem.

"To be sure, 97 quadrillion sounds like a number too big to comprehend. It's 100 times greater than the number of ants on Earth, but 10 times less than the number of molecules in a snowflake, if that helps to put it into perspective."

"Not really," I complained.

Undaunted, he went to work. With a zap from his bony index finger, a large shining yellow orb appeared in the sky overhead. It radiated intense heat. "Behold, the sun!" he proudly announced. Next, he drew a circle in the air with the fingers of both hands. A small Earth appeared, suspended in the air in front of me. It looked so real I was afraid to touch it, considering the havoc I might create for all the tiny people and animals and forests if I did.

"Now, the Earth travels around the sun once in every 365.25 days. To do this it has to travel at 18.51 miles per second. If you laid matches end to end, there would be 551,894 kitchen matches in 18.51 miles. So, here's the question: if you set out kitchen matches behind the Earth as it traveled around the sun—at the rate of 551,894 matches per second—would you run out of matches before the Earth rammed into the beginning of the 584 million-mile chain of matches you carefully started a year before?" Zed's illusory Earth began to move in a graceful arc around his homemade sun, as a stream of infinitesimally tiny matches trailed behind it.

At over a half-million matches per second, it seemed like you'd *have* to

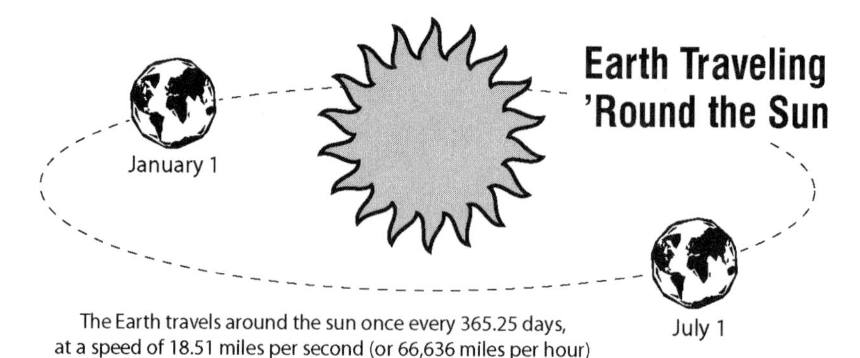

The Earth travels around the sun once every 365.25 days, at a speed of 18.51 miles per second (or 66,636 miles per hour)

run out of matches, but something made me say, "No, I don't think so."

"Indeed you wouldn't!" he proclaimed, "not after one year, or 100 years or even 1,000 years. In fact, if you had begun setting out 551,894 matches per second in 3500 B.C., just about the time your ancestors were toying with the idea of writing—and, alas, trying to domesticate the cat; an exercise in futility, if you ask me—you would just now start running short on matches."

I watched in awe as the Earth became a blue-green blur, with an ever-thickening reddish-yellow tail. Abruptly it stopped and, with a wave of Zed's hand, disappeared.

"Great Scott!" I exclaimed.

Zed patted me on the shoulder, and replied, "Well said, lad, well said."

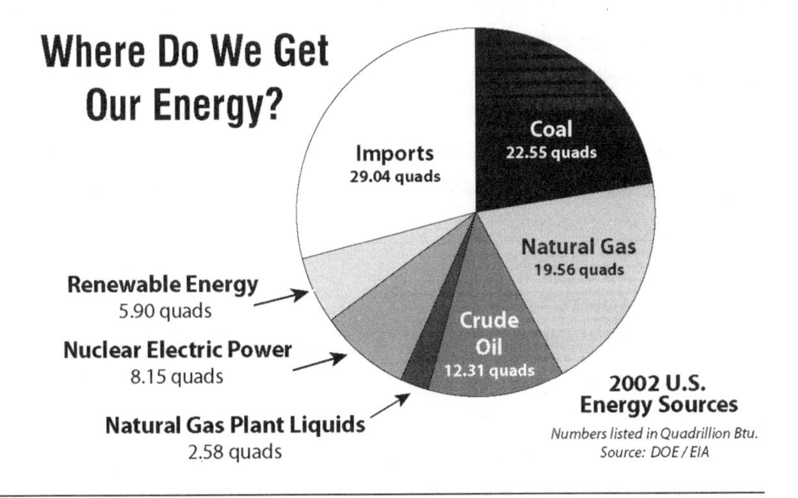

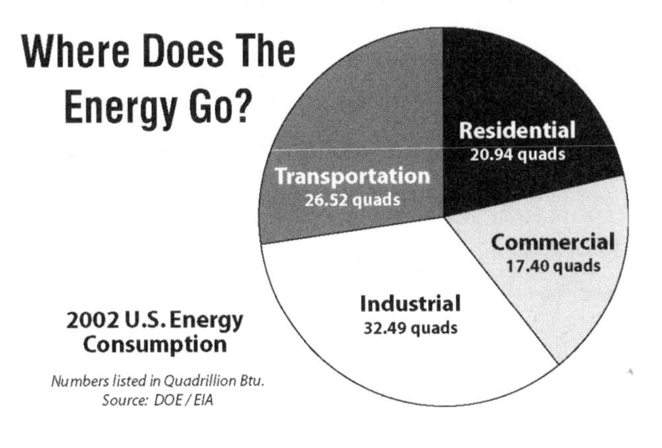

In 2002 energy consumption in the U.S. exceeded our production and we imported over 29 quadrillion Btu of energy (primarily crude oil).

technistoff - 7

While it was obvious Zed was having a great time delivering this lesson, I myself overheated more than a few brain cells checking his numbers.

numbers and more

The enormous figure of 97 quadrillion Btu for the year 2002 was confirmed by our government's own number-crunchers. I found it at *www.eere.energy.gov*. A more accurate number is 97,590,000,000,000,000—a little bigger than Zed's estimate. This website also provided the percentage of Btus used by each of the four energy sectors.

In *The Sizesaurus* I learned that a kitchen match does, indeed, deliver one Btu of heat, but that's only when it's used to light a candle, or a stove. Let it burn to the end and it will give off nearly two and half times that much heat, if you consider a 5-foot cube of matches (125 cubic feet) about equal to a cord of ponderosa pine firewood (128 cubic feet), the latter providing 17,600,000 Btu of heat. Still, it made a great analogy, and I can't blame Zed for wanting to use it.

Zed's calculations for the depth of matches in Rhode Island are correct, if you assume a match to be 1/8 inch x 1/8 inch x 2-1/8 inches *[3.175mm x 3.175mm x 53.96 mm]*, and Rhode Island to be 1,214 square miles *[3,145 sq km]*.

My home solar and wind system does produce, on average, around 6 kWh per day. Considering that most homes use four times that much electricity, Zed's guess that conservation measures would save other homeowners 6 kWh/day is extremely conservative.

I verified his claim that 70 percent of the energy used to produce electricity is wasted, by visiting the *www.flexibleenergy.com* website. Flexible Energy's figures were compiled using DOE data. I should point out here that the Energy Information Administration suggests a figure closer to 66 percent—around 10,200 Btu of fuel consumed per each kWh of electricity produced—but this figure does not incorporate the 10 to 12 percent loss in transmission. Flexible Energy, by contrast, takes the amount of fuel consumed by electrical utilities to make electricity—40 quads in 2002—and compares it to the amount of electricity actually used by consumers—around 12 quads. Since most of us don't plug our toasters into sockets mounted on the walls of the local power plant, I feel the larger figure to be more accurate.

inefficiencies of creating & moving electricity

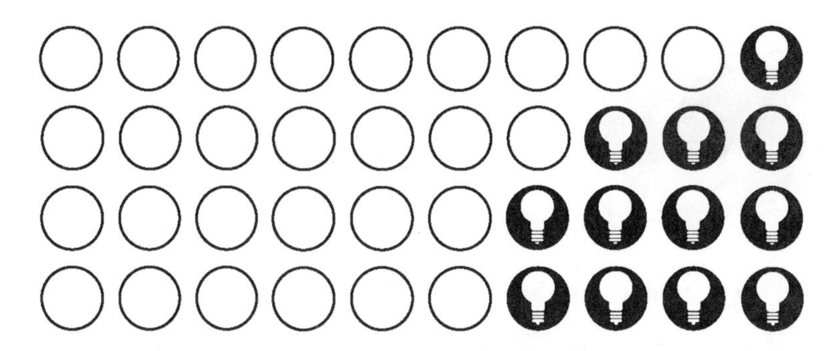

Inefficiencies of Creating & Moving Electricity

In 2000, approximately 40 quadrillion Btu of energy were consumed by electrical power generators in the United States, but only 12 quads-worth of electricity made it to the consumer. *Source: Flexible Energy, Inc. / DOE-EAI.*

To verify the specific gravities (densities) of broken and solid bituminous coal, and of oil, I used the figures of 833 kg/m^3, 1,346 kg/m^3 and 881 kg/m^3, respectively. This information is available at *www.simetric.co.uk/si_materials. htm*. The depths Zed calculated these materials would cover Rhode Island are correct, using energy values of 11,700 Btu/lb. *[27,215 kj/kg]* for bituminous coal, and 18,800 Btu/lb. *[43,730 kj/kg]* for #4 crude oil, as I found at *http://home.att.net/~alt.hvac/fuels.htm*. (For the record, the specs for anthracite coal are: broken 1,105 kg/m^3; solid 1,506 kg/m^3; and 12,700 Btu/lb *[29,540 kj/kg]*).

Zed's off-the-cuff estimates for butter and maple syrup were easily verified with a trip to the refrigerator to read the food labels.

Zed used a figure of 61,095 Btu/lb. *[142,100 kj/kg]* of hydrogen (*www.engineeringtool box.com*) and a density for liquid hydrogen of 0.07 gm/cm^3, which I discovered at *http://hyperphysics.phy-astr.gsu.edu/hbase/pertab/h.html*. To verify that a 5-gallon bucket of liquid hydrogen would weigh about the same as a feather pillow (around 2.9 pounds), I weighed my pillow. Not having a bucket of liquid hydrogen, I accepted the figure of 0.07gm/cm^3 as gospel.

The discharge rate of 600,000 cubic feet per second Zed gave for the Mississippi River is rounded up from the more accepted figure of 593,000 cubic feet per second *[16,794 cubic meters per second]*, as I discovered at *http://outreach.missouri.edu/mowin/ Rivers2/missriver.html*, as well as several other sites.

The number of ants on Earth and the number of molecules in a snowflake were found in *The Sizesaurus*.

All the conversion factors I used to do the math were found in the *Pocket Ref.* This includes the theoretical value of 3,410 Btu per kWh, which differs slightly from 3,412 Btu per kWh suggested by the IEA, and others. Where Zed's conversion factors came from, I have no idea. Nor do I have a clue how he did his calculations so quickly. I labored over mine on a Texas Instruments TI-35 PLUS pocket calculator.

NOTE: As you delve into energy matters you will encounter a number of confusing units. This can't be helped. Most of the world has abandoned the units we in the U.S. cling to so dearly, and the scientific community, being global in it's endeavors, has followed suit.

kilojoules & exajoules

Specifically, you will often see heat expressed in **kilojoules** (kj, or 1,000 joules) rather than Btus. Don't despair; unless you find yourself doing precise calculations, you can pretty much think of a kilojoule as being equal to a Btu. (The actual conversion is: 1 Btu = 1.054 kilojoules, or 1,054.35 joules.)

Likewise, while we in the U.S. like to think in **Quads**, almost everyone else thinks in **Exajoules**. Again, the units are close. An Exajoule = 10^{18} joules, while a Quad = 10^{15} Btu. Simply put, a Quad = 1.054 Exajoules, or an Exajoule = 0.948 Quads. Close enough for most mental calculations, at any rate.

HHV versus LHV

Another area of confusion I have attempted to avoid (but which you will doubtless encounter in other literature) is the difference between the **Higher Heating Value (HHV)** and **Lower Heating Value (LHV)** of various fuels. For example, the HHV for hydrogen is 61,095 Btu/lb. *[142,100 kj/kg]*; the LHV is 51,630 Btu/lb. *[120,000 kj/kg]*.

The difference lies in the way the measurement is made. The LHV is simply the amount of heat given off during combustion, while the HHV is the LHV **plus** the energy recovered as the products of combustion cool and condense. This includes the water vapor present in the fuel, as well as water that is formed from hydrogen liberated during combustion.

In most common systems (car engines, furnaces, stoves, etc.) only the LHV applies, since no one bothers to capture the energy of combustion condensates. Other far more sophisticated systems—as we are soon to encounter—use every available Btu (or kj), and in these cases the HHV comes into play.

book two

★ ★

WHERE IN THE WORLD IS THE HYDROGEN?

Methanol
Renewables Coal
Ethanol Natural Gas
Nuclear Steam Reforming
Electrolysis Gasification
Thermal Water Splitting
Biological Thermochemical
Photoelectrochemical
Natural Gas Splitting
Steam Electrolysis
Biomass

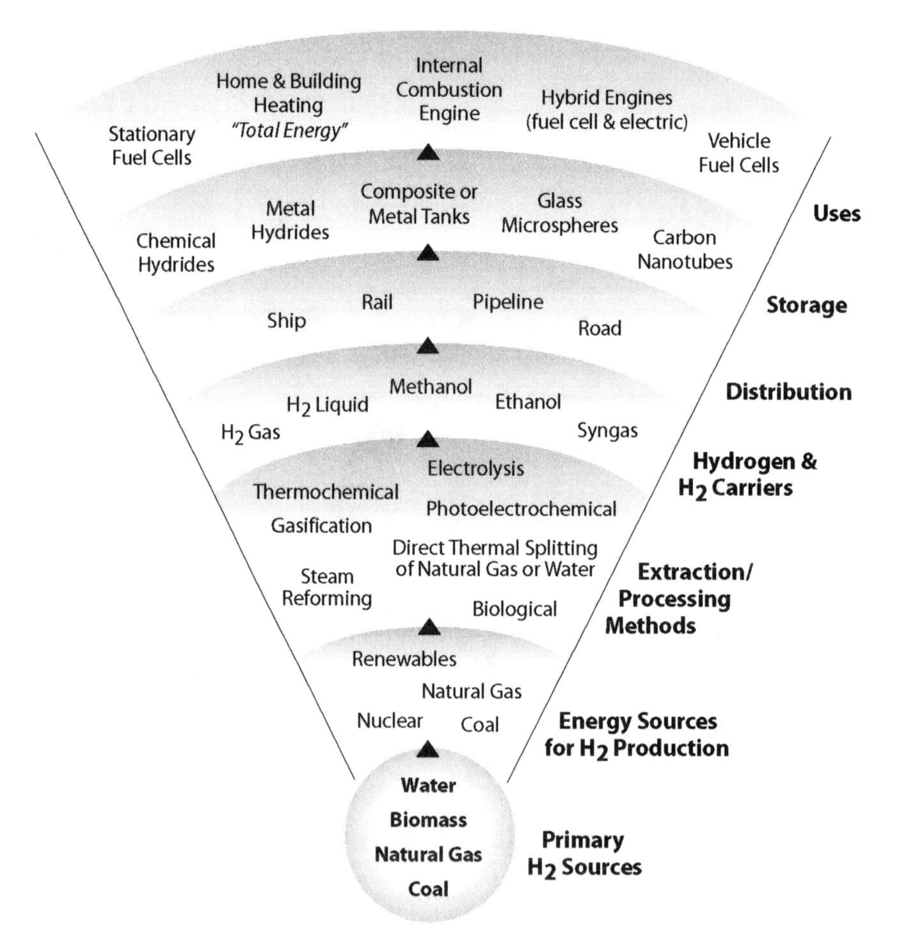

Moving Towards a Hydrogen Economy

chapter 8

Splitting Water: It's a Matter of Perspective

The food was getting better, I had to admit. The breakfast I found waiting for me the next morning bore far more resemblance to typical ranching-country vittles than the marooned-on-a-tropical-island fare I'd been getting. There was actually a pair of real eggs on my plate! They were nestled next to some soy concoction that was meant to resemble chopped steak, but did so only in a rudimentary way. Oh, well. I was making progress so I decided not to complain. As always, the coffee was superb and the fresh fruit—now thankfully relegated to the role of side dish—was as succulent as ever.

Zed was waiting for me as I stepped out into the day. His clothing was unchanged (as was mine), except that his T-shirt now said: *Fossil Fuels are Fossils at the Wasserstoff Farm*, and he was wearing a ball cap sewn from festively colored blue, red and green triangles. The words *Wasserstoff World* were embroidered on the front. It was the most color I'd ever seen Zed wear, and it looked more than a little out of place on the long-haired old Druid.

"Nice hat," I lied.

"Why, thank you. I thought it would go well with the day's activities."

"Activities?" I said, "as in 'doing something other than simply observing lessons'?"

"Correct," he affirmed.

This was beginning to sound exciting. I pumped him for answers, asking, "Really, Zed? So, where are we going? And once we get there, what are we gonna do?"

"Honestly," he said, with an audible sigh, "you're worse than an attention-starved five-year-old." His eyes softened when he saw my look of dejection, and with a broad smile, he asked, "How about we go to a carnival?"

"A carnival? Like with Ferris wheels and loop 'd loops?"

"Uh...something like that, I suppose," he replied, a bit evasively. "Actually, I was hoping to try out one of the attractions at the newly-completed Wasserstoff World theme park."

Try out? Theme park? Why did I suddenly feel like a test pilot?

"Anyway," he continued, "I'm sure you'll enjoy yourself."

I wasn't sure of anything anymore, but I followed Zed into a small building several dozen yards behind the holosium. "It's not exactly part of the Wasserstoff Farm proper; it's too big to fit inside. We'll have to take a little trip. Follow me; this is the easiest way to get there," he told me, directing me into a dark and windowless room.

How could *anything* be too big to fit inside the Wasserstoff Farm, I wondered? And how could we possibly get anywhere from inside a cramped little building? The answer to the first question would be a long time in coming. The answer to the second question, however, came like a bolt of lightning, for no sooner had I set foot inside Zed's mysterious little room than I felt like I'd walked into an electrified chain-link fence wearing a copper suit. My hair stood on end and a sharp, biting tingle coursed like a shock wave through every nerve in my body.

Mercifully, the feeling was short-lived. Before I even had time to scream I found myself standing in a very peculiar place. An instant later Zed appeared beside me, his long, white hair sticking straight out past his shoulders from beneath his ball cap. "I need to work the bugs out of that thing," he said, smoothing down his flowing locks.

"Maybe you should adjust the flux capacitor," I suggested, recalling the old *Back to the Future* movies.

Zed shook his head. "I already tried that. I think the problem lies more with the plasma field modulator. Whatever. I'll fuss with it later. Today you're going to learn about one of the more straightforward ways to liberate hydrogen from water."

Water? I scanned the surroundings. The ground beneath our feet was hard-packed red sand. It looked as if it hadn't seen a drop of water for a thousand years. If not for the fact that we were still breathing, I'd say we were on Mars. In the distance the perfect nothingness before me was broken up by what appeared to be a colossal chemistry lab, rising from the desert floor to several hundred feet into the air.

I no longer felt like a test pilot. Now I was a lab rat. "Where are we, anyway?" I asked.

"Where? Why do you always have to know where you are? We're here, in a desert."

"I know, Zed, but *which one?*"

"Which one? Um, let's see...oh, I forget. Why in the world would it matter?" Before I could say any more he set off briskly across the parched sand. Reluctantly, I followed, hoping conversation might calm my nerves.

"I need to get something straight, Zed. The universe is mostly hydrogen, right?"

"That's correct," he answered, without slacking off his pace.

"Yet there's very little in our atmosphere that isn't bound up in compounds, like water vapor, methane or hydrogen sulfide?"

"For all intents and purposes, there's none," he answered. "That's because it all leaked out, eons ago."

"Leaked out?"

"Right. Lightweight molecules travel faster than heavier ones. At room temperature hydrogen travels at an average speed of around 420 miles per hour. Heat up the air and it travels even faster. In former times when the air was a great deal hotter than today, hydrogen moved much faster. Given enough high-energy collisions with surrounding gas molecules, almost all the unbound hydrogen that was formerly in our atmosphere eventually reached escape velocity and vanished into space."

"Escape velocity? What's that?"

"It's the velocity at which something—be it a rocket or an atom—has to be moving away from the Earth to escape its gravitational field; about 25,000 miles per hour."

"Wow!" I exclaimed, though not so much from what Zed had just said, as from what now loomed before me. Rising from the desert floor and reaching toward the sky were two transparent tubes, each several tens of feet in diameter. Centered in each, I could see a tall vertical shaft, or rod, of shiny metal. Curiously, a giant, odd-shaped balloon was draped loosely down the backside of the structure, its two openings tightly stretched over outlets affixed to the tops of the two towers.

Offset toward the bottom, the towers were joined by a transparent horizontal tube, or corridor, which allowed the fluids to mix within the towers, and gave the giant apparatus the appearance of an H. From the middle of the H there rose yet another transparent tower, far narrower than the two side ones, with a bulbous opening on top. A platform surrounded this opening, and attached to the platform was an elevator shaft descending to the ground.

Suddenly, my knees felt weak.

"Well, then, here we are," Zed said. If he had any misgivings, they were not betrayed by his voice. "Let's get started, shall we?"

The elevator was just big enough for the two of us. On the long ride to the top, Zed explained what we were doing. "This structure we're scaling is called a 'Hoffman apparatus'. It's nothing like a commercial electrolyzer; it's more a laboratory curiosity. However, it's perfect for our purposes, since all that's important right now is for you to know how the principle of electrolysis works. I think when Wasserstoff World opens we'll call it 'The Wasser Kleaver'."

I didn't like the sound of it, and I told him so. "You worry too much," he told me, as the elevator ground to a halt.

We stepped out onto the large platform surrounding the top of the opening, which contained a lake at least 40 feet across. A small two-man submarine was moored to the side. It was bright yellow. "Ticket, please!" Zed said, directing me into my seat.

"I left it in my other pants. You'll have to find another customer," I said, hopefully.

Undaunted, Zed slipped his hand around behind me and extracted a ticket from my back pocket. *Admit One to The Wasser Kleaver*, it read. "You're in luck," he announced, brandishing the ticket like a hand grenade. My heart sank to my nether regions as I stepped into the sub.

The ride down the inlet into the Hoffman apparatus was almost relaxing. The sub dropped gently and evenly. Zed deftly worked the controls as I gazed out the large, curved Plexiglas windows. It would've been nice to see some fish, I thought, before realizing how impossible that would be.

By the time we reached the horizontal corridor connecting the two towers, I was ready for something to happen. It didn't take long. "As I'm sure you know by now, all molecular bonds are electromagnetic in nature. There is very little difference between the positive and negative poles of a battery, and the opposing charges of the proton and the electron.

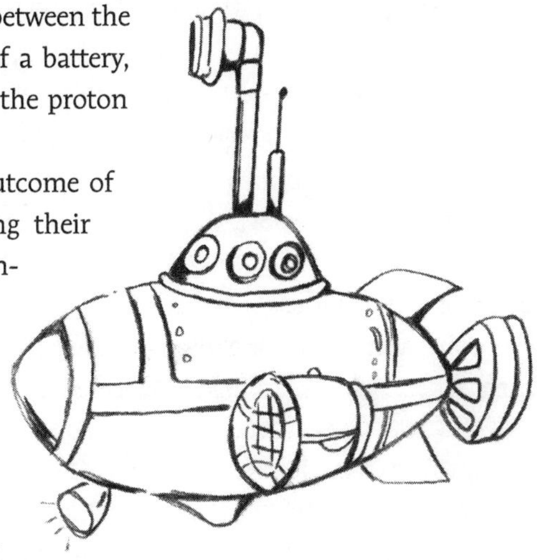

"Water is the natural outcome of hydrogen and oxygen seeking their lowest energy states," he continued. "Hydrogen and oxygen give up a great deal of energy to become water, and it takes a lot of energy to separate them again." He turned a knob on the sub's console and instantly I could feel a slight, but noticeable tingle in the air inside the sub. More disturbing, however, was the fact that the sub seemed to be receding from the sides of the tube at an alarming rate.

"*What's happening?!*" I exclaimed.

"We're getting smaller," he said, evenly.

"How *much* smaller, Zed?"

"Several orders of magnitude, actually. But don't worry—it's temporary."

So is life, I reminded myself, as I gripped the arms of my seat and watched with equal measures of dread and fascination as the constituents of the fluid in which we were suspended became visible.

Soon our sub was just a little larger than the individual water molecules around us, if indeed that's what the ghostly shapes careening off one another just outside the sub were.

Each molecule looked like a pair of fuzzy, vibrating clouds merged with one much larger cloud. Things were moving within the clouds, far too quickly for the eye to capture. The nervous molecules formed a constantly shifting matrix around the sub, like a loosely-bound lattice that was constantly in motion. The molecules were energized by the slight charge flowing through the water and their motion jostled the sub like a cork in a pond. In places the molecules swirled in giant eddies, and from time to time as we hovered about, we would get caught up in currents—caused by uneven heating of the water by sunlight, I surmised—and swept swiftly around the enclosure. I fought to hold onto my breakfast.

"Let's turn up the juice, shall we?" Zed suggested. Before I could object, he cranked the knob hard to the right. Things changed in a hurry. "There is now a stream of energetic electrons flowing from the cathode in the tower to our right, to the positively charged anode in the tower to our left."

How could he know what was where? More importantly, how did he intend to get us out of this mess? Though I couldn't see the electrons, I could feel them pinging off the sides of the sub. More distressing, the water molecules were beginning to break apart, the oxygen atoms flying off in all directions. It was like trying to navigate a starship through an asteroid field.

Zed was unperturbed. "If you'd quit fidgeting and pay attention," he admonished, "you'd notice something very peculiar happening around you."

He was right. I don't know what I thought a water molecule was supposed to look like when it disintegrated, but this wasn't it. Instead of the two small fuzzy balls imbedded in the big one simply floating away, the smaller balls simply popped out of existence. "Zed! What's happening to the hydrogen atoms?" I asked. "They're disappearing!"

"No they aren't. You just can't see them."

"Why not?" I wondered.

"Can you see a beach ball from ten miles away?" he smugly asked.

Beach ball? Of course! Now I understood. The oxygen atoms were holding onto the two electrons provided by the hydrogen atoms. Without an electron cloud surrounding it, the hydrogen nucleus—a single naked proton—was a small thing, indeed.

"Don't worry. They'll find electrons, soon enough. Let's take a spin over to the cathode and catch the action." With that, Zed spun the sub around, setting me back in my seat as he sped off to parts unknown.

Everything was a blur as the sub rocked and swooned from atomic-scale impacts, but in a moment we arrived at our destination.

Once we were within sight of the huge cathode, I could see hydrogen atoms puffing into existence like popcorn blossoming from invisible kernels. Quickly they joined together in pairs and rose toward the top of the tower, out of sight. The closer we moved toward the cathode, the more intense the action. By the time we reached the cathode, it was pure pandemonium. Viewed on a larger scale, it must surely have looked like water boiling.

"This is where the action is," Zed said, looking with wonder at the sight before us. "The protons flock to where the electrons are, like moths to the flame. Isn't it beautiful, all those tiny protons, luring electrons to them, before mating and floating toward the sky?"

It was.

"What about the oxygen?" I asked. "What's happening on the other side of The Wasser Kleaver?"

Zed pulled a slick kitty with the sub, and I held on tight for a wild ride. In a moment we were there. Visually it was not nearly as exciting as watching hydrogen materialize from the void, but, considering the extra size of the oxygen atom, the action was even more intense. Like a wild feeding frenzy in reverse, the oxygen atoms couldn't wait to give up the two extra electrons and become electrically neutral again. Because each oxygen atom still had six electrons after giving up the extra two, their appearance was hardly changed.

Once unburdened of their invisible loads, the oxygen atoms paired-up and floated out of solution. "Why does it float to the top?" I wondered, "isn't O_2 heavier than water?"

"Of course it is. But it's not bipolar like water is. Without hydrogen bonds, there's no electrical attraction between molecules. It's free to go where it pleases." He said this with a distant, admiring gaze. "Seen enough?"

Zed sez

Hydrogen and oxygen give up a great deal of energy to become water, and it takes a lot of energy to separate them again.

he asked. I nodded. Zed worked the controls and steered us toward the central tube where we began to rise, growing in size as we did. By the time we reached the top, we had grown to normal size.

Stepping out of the sub on shaky legs, I noticed that the large balloon stretched over the tops of the towers still lay flaccid against the side of the Hoffman apparatus. "That's going to hold all the O_2 and H_2 we just made. Two hydrogen molecules for every oxygen. I had to wait until we were out of the sub to fill it, otherwise the water in the central column would have fallen so low we wouldn't have been able to get out. Care to see what happens when I open the valves and light a match?" He held a wooden match in his hand with his thumbnail poised against the tip.

"Uh, Zed...do you suppose we could take the elevator down, first?"

Taking a bearing on where we were, he said, "Sure. Not a bad idea."

Once on terra firma, I looked up at the giant towers. Both held less water than before, though the cathode side, where the hydrogen had collected, held half as much water as the oxygen side. I heard Zed saying, "An equal number of molecules of different gases will occupy the same volume, no matter what they weigh. As you see, twice as many hydrogen molecules displace twice as much water. If you don't believe me, ask Avogadro—he's the one who figured it out."

I decided to take Zed's word for it. "So how much hydrogen did we produce, anyway?" I asked.

"Well, it looks like we broke down about 20 tons of water, which would liberate around 4,500 pounds of hydrogen. At 60,000 Btu per pound, that's about 270 million Btu. Enough heat energy to run a really efficient fuel-cell car for about 150,000 miles."

"Impressive!" I exclaimed. "And how much electricity did we use?"

"Well, now...that's the rub," he said, eyeing the meter beside the elevator. "We used 117,000 kWh of electricity. Probably enough to run the same car for over 220,000 miles, providing the electricity was transformed into heat energy with 100 percent efficiency."

"*What?*" Forget the mileage; I was incensed. I screamed, "I was in a water tank with *that* much juice going through it?"

"I didn't think it was anything you'd want to know beforehand," Zed pointed out.

"I'd never have gotten in!" I exclaimed.

"Exactly. And just think what you would've missed. I swear, you do a man a favor and he turns on you. Why do I even try...?"

Now I felt bad. "Wait...Zed...it's okay. I suppose. No harm done, and all that. And I *did* learn a lot, after all."

"I would hope so," he said, eyeing me cautiously.

I swept my eyes around the grounds, looking for power lines. There were none. "Where did all that electricity come from, anyway?"

"Oh, you'll find out, soon enough. Anyway, the lesson's not over. Are you going to stick around for the finale, or do you want to run home with your tail between your legs?"

"Home? I don't even know where home is!"

"Then I'll take that as a 'yes'."

He flipped a switch on the side of the elevator, and a loud, ominous *hissss* issued from the tops of the towers as the balloon began to fill. Zed stepped around to the back of the elevator and returned with what looked like an English longbow and an arrow with a rag tied around it. It smelled like it had been dipped in kerosene. "Fossil fuels are good for some things," he said, raking his thumbnail over the match and lighting the rag.

He nocked the arrow and drew back the bow until his hand was nestled against his jaw. Taking aim at the now-full balloon several stories overhead, he released the arrow. The instant it plunged into the balloon I braced myself for a deafening explosion; instead, it was more of a colossal *whuuuumph!,* as from an enormous *implosion*. And the huge fireball I expected to see engulfing the top of the Hoffman apparatus was but a faint, bluish ripple of heat, rising in the air—followed by a rush of air toward the invisible flame that nearly knocked me off my feet. Then, for an instant, it rained buckets. Literally. It was like a tidal wave from heaven. Dripping wet, I glared at Zed in disbelief.

He asked, "So...what's it like to bathe in 12 years of car exhaust? Is it worth the cost of the ticket?"

I shivered in silence.

technistoff - 8

Just when I thought things couldn't get any more bizarre…

As much as I'd like to be able to explain what just happened, I can't. Any attempt to understand the technical hurdles Zed had to overcome to build a skyscraper-sized Hoffman apparatus would be folly. Likewise for the incredibly shrinking yellow submarine. I suspect it was a trick involving extra dimensions, but the bare notion reveals nothing useful of the working concept, so I'll leave it at that.

In any event, it has always been beyond the scope of the *Technistoff* sections to illuminate *how* Zed does what he does. It's enough just to affirm the veracity of what he *says*.

Isaac Asimov speaks of the average speed of the hydrogen molecule in *Building Blocks of the Universe*. He gives it at 7 miles a minute, which works out to 420 miles per hour *[675 km/hr]*. The escape velocity for Earth and all the other planets *[25,000 mph or 40,225 km/hr]* is given in *The Handy Physics Answer Book*. The fact that any hydrogen which didn't form compounds with other atoms escaped the Earth's gravitational field is self-evident.

No one I know of (other than me and Zed) has ever viewed the electrolysis of water to liberate hydrogen and oxygen on the atomic scale, so the exact

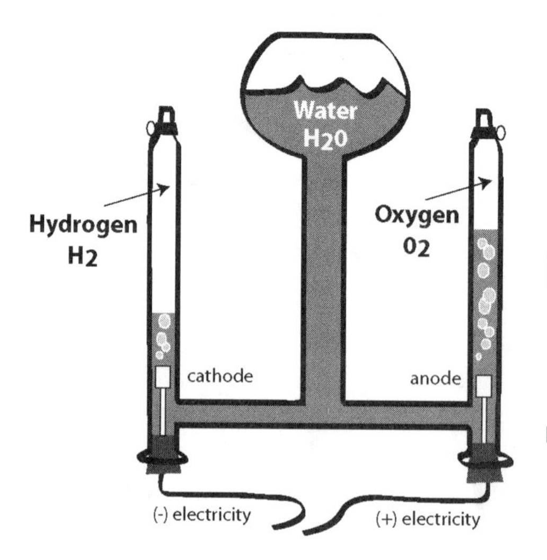

Electrolysis

A Hoffman apparatus demonstrates how electricity is used to split water into hydrogen and oxygen.

order of things is a little unclear. All I can say is, I know what I saw. The protons quickly migrated to the negative cathode in hopes of finding electrons to replace the ones stolen from them by the oxygen. The electrons stayed with the oxygen ions, since oxygen's ability to take them exceeded hydrogen's ability to keep them. Having a net charge of -2, they were drawn to the positive anode.

electrolysis of water

As Zed said, 20 tons *[18.14 tonnes]* of water—88.8 percent of which is oxygen—does contain about 4,500 pounds *[2,041 kg]* of hydrogen, or right around 270 million Btu. Divide this by 125,000 Btu per gallon of gasoline, and you get 2,160 gallons which, at 70 miles per gallon—the "gasoline-equivalent" miles that hydrogen fuel cell cars are one day expected to achieve—will propel a car for 151,200 miles.

Bearing in mind that Zed's monstrous electrolyzer was set up more for show than production—and probably operating at less than 70 percent efficiency—the 117,000 kWh of electricity it used during our wild ride would be about right, using a conversion factor of around 3,410 Btu per kWh.

Although Zed never mentioned it—caught up, as he was, in the sheer grandeur of his theme park attraction—there must certainly have been electrolytic ions dissolved in the water. Water, by itself, is a poor conductor of electricity. The addition of sulfuric acid (H_2SO_4) or potassium hydroxide (KOH) to the water in an electrolyzer greatly increases the efficiency of the process, as does pressure and heat. This, of course, would have led to a complicated series of reactions involving the rapid making and breaking of bonds and the formation

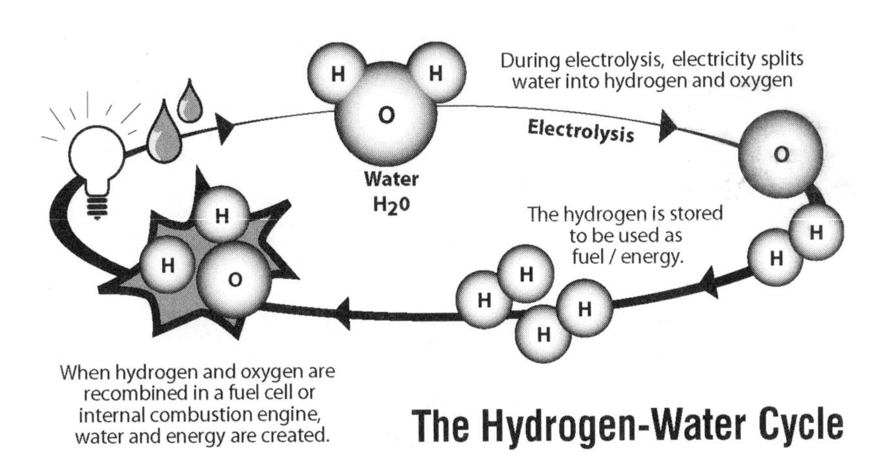

The Hydrogen-Water Cycle

of several transitory intermediate compounds. Needless to say, it would've made the whole demonstration incomprehensible from our point of view. Ergo, Zed kept it simple.

I should point out here that the 2:1 hydrogen-to-oxygen gas mixture inside the balloon—often referred to as Brown's Gas—is not usually the desired product of electrolysis. Rather, the whole point of electrolysis is to *separate* hydrogen from oxygen, at least until we've used the hydrogen—in a motor or a fuel cell, for example—at which point it is indeed recombined with atmospheric oxygen to make water. But these are things Zed must be saving for later lessons. The point of *today's* lesson was to demonstrate the amount of energy available in such a large volume of hydrogen gas.

There was no fireball because hydrogen burns with a nearly-invisible flame. And, because the ignition of Brown's Gas results in an instant and dramatic *reduction* in volume and pressure, the mixture was seen to implode, rather than explode.

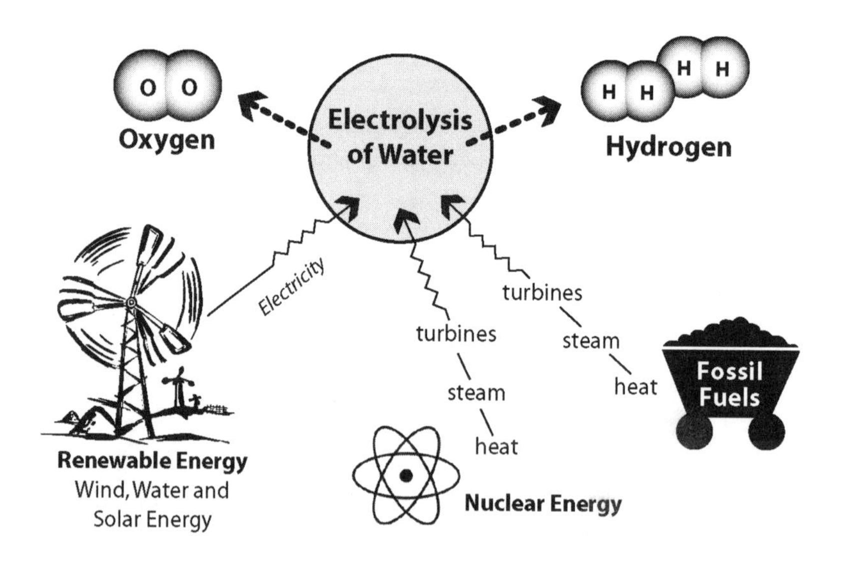

Sources of Electricity for Electrolysis

Renewable energy sources, such as wind farms, solar arrays, and hydro-electric dams are the ideal sources of electricity for electrolyzers, but the more common sources of fossil fuels and nuclear energy provide the majority of electricity for this not-yet-popular method of liberating hydrogen.

In the real world, electrolysis is conceptually simple, and relatively uncomplicated on the practical side. Best of all, electrolysis produces hydrogen of very high purity, cutting down the number of steps needed to produce a product clean enough to be used in fuel cells. There are several manufacturers currently producing high-volume electrolyzers, utilizing a number of technologies. Perhaps the most promising of these is the PEM (proton exchange membrane) electrolyzer, which uses a porous, conductive membrane, rather than electrolytic chemicals, to maximize the flow of current from anode to the cathode. These electrolyzers operate in much the same way as a PEM fuel cell but in reverse, and certain models can, indeed, perform both electrolyzer/fuel cell functions.

purity of hydrogen

The best industrial electrolyzers operate at near 90 percent efficiency, and even greater efficiencies (though probably not much greater) may be achieved in the future. In fact, the technology currently exists to continually produce enough hydrogen by electrolysis to run all the world's cars, trucks and buses—if not for three nagging caveats. These are: 1) the cost of producing the electricity to run the electrolyzers, since most of our electrical generation facilities are woefully inefficient; 2) the fact that the fossil fuels used to create the electricity needed to produce hydrogen are more polluting than simply running cars on gasoline; and 3) the obvious fact that there are no cars currently in mass production with the capabilities to use hydrogen as fuel.

electrolysis obstacles

This latter concern is the least of the problem. The only real obstacle to mass-producing efficient hydrogen-fueled cars is not so much technological—it's simple dollars and cents. There is little point in manufacturing a fleet of very expensive hydrogen-fueled vehicles if there is no place to buy hydrogen produced under environmentally-responsible conditions.

Ultimately, then, the future of clean, efficient, non-polluting transportation may well depend on clean, efficient, non-polluting sources of electricity.

Of course, as Zed is quick to point out, nothing is ever quite that simple…

DO-IT-YOURSELF ELECTROLYSIS EXPERIMENT

It's often said that seeing is believing. Whenever I hear this old dictum I recall the horror I felt as a naïve six-year-old, the day my parents took me to a magic show and I watched an obviously-deranged magician saw a perfectly healthy woman in half. Never mind that not a drop of blood was spilled, or that the magician somehow managed to put the woman—inexplicably undaunted throughout the entire escapade—back together again with even greater ease than she came apart. For a time that poor woman was in two pieces. I know, because I saw it.

Obviously seeing isn't always enough. Sometimes you just have to saw the woman in half yourself. It's the only way to truly appreciate the complexity of certain things. Like electrolysis. Just because everyone says you can split water into oxygen and hydrogen with a couple of volts of direct current doesn't mean it's true; it might all be smoke and mirrors. The only way you'll know for sure is to do it yourself.

If you're willing to give it a try, here's what you'll need:

- ★ One 6-volt lantern battery (like the one in the big flashlight you never use)
- ★ Two pencil leads (the heavy kind used as replacement leads for retractable construction pencils work best)
- ★ Two short pieces of insulated wire, 14 to 18 gauge
- ★ Two small alligator clips
- ★ A small dish (glass or ceramic) filled with warm salt water

Having gathered all these things together, you simply need to strip both ends of the wires, attach the alligator clips to one end of each wire, and wrap each of the other ends around the pigtail terminals on the battery. Then secure a pencil lead in each of the alligator clips and carefully lower your two graphite electrodes into the water.

There should be an immediate and robust reaction on the negative (cathode) side as the H+ ions migrate there in search of electrons. The positive (anode) reaction will take a little longer to get going, but after a few seconds you'll see oxygen bubbles begin to form on the graphite surface. The anode reaction will be about half as vigorous as the cathode reaction, since only half as much oxygen gas is formed.

To convince yourself that you are really seeing the formation of H_2 and O_2 gas, you can invert a test tube (or even a pill bottle) over the water near the cathode. It will take some patience to capture enough hydrogen gas to combust when you light it with a match, however. A quicker and easier way is to hold the match near the cathode and observe as the flame grows and crackles, like light rain falling on a newly-fed campfire. The crackling is caused by water vapor forming within the flame, cooling the fire as it consumes fresh new fuel.

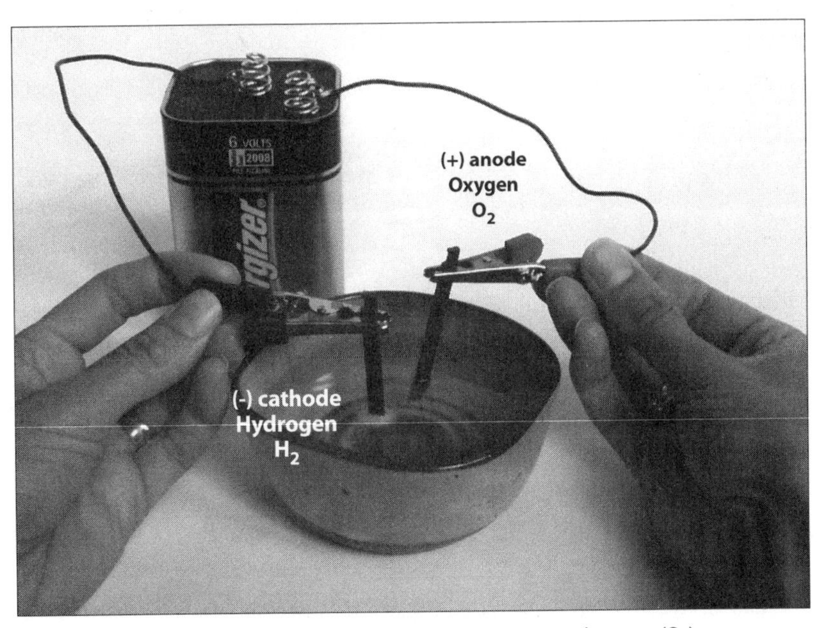

A cheap and simple way to split water (H_20) into hydrogen (H_2) and oxygen (O_2).

For those questioning the wisdom of sticking charged electrodes into a vessel of water, don't worry—just be sure not to touch the two electrodes or the alligator clips together. Beyond that, it takes a lot of work to shock yourself with DC electricity—I know; I've tried about everything. It's the reason no one ever attempted to build a DC electric chair.

This being said, I would caution you gung-ho types to think twice before hauling your DC arc welders upstairs to your bathtubs. All things have their limits.

chapter 9

How to Perform an Electrolyzer Bypass

I was surprised when Zed informed me that the rest of the lesson on splitting water to make hydrogen would be taking place back at the Wasserstoff Farm. And, I have to admit, maybe even a little disappointed. Although Zed's attitude toward safety bordered on utter disregard, he kept things interesting. And there were a lot of interesting-looking attractions waiting to be explored at Wasserstoff World.

"Later," he said, as we passed through the Dimensional Compressor—the strange doorway that somehow temporarily subdued ("collapsed" was the word Zed used) one of the three spatial dimensions, and always caused Zed to have a very bad hair day. "Everything in its time. For now, you have to understand that huge electrolyzers aren't the only way to split water into hydrogen and oxygen.

"The beauty of electrolyzers lies in the fact that they're simple, the technology is currently available, and they can be scaled up in modular units to any size that's practical for the application. The downside is that, by and large, the electricity needed for electrolyzers to work their magic is, in our current primitive technological state, dirty and wasteful."

He stopped walking and raked me with a critical eye. "The people who harangue about the wasteful part of it are being unfair, of course. There are gross inefficiencies inherent in any system that produces electricity. Commercial flat-plat solar modules are currently only 12–14 percent efficient, and wind systems are generally less than 40 percent. And that's before the power is moved—transmitted—from the source to the point of use. Wastefulness is a fact of life. But dirty doesn't have to be. We could learn to live with less-than-perfect efficiencies if our failings weren't making such a mess of the atmosphere. Green plants, after all, have done well for hundreds of millions of years by utilizing a mere 3 percent of the sunlight that falls on them. Of course, what they manage to do with that 3 percent is so mind-bogglingly efficient it's beyond mere human comprehension, but that's another story...

"Anyway," he said, as he resumed his energetic pace, "we'll bring the efficiency issue into perspective in due course. For now I want to show you a few of the experiments we've been working on here."

"We?" The word caught my attention. I asked, "Who's *we*? I haven't seen another soul since I got here!"

"Yes, well, that's to be expected," he answered, skirting the issue. "Look—here we are. Isn't it beautiful?"

He was indicating a large tank of water with mats of green gunk floating on top. "Looks like pond scum," I answered honestly.

"And so it is," he admitted, "but not just any pedestrian variety of pond scum. This is a very talented species of pon...uh, algae, that is. You could run a car on what this stuff exhales."

"Is that so?" I thought about it for a minute, then said, "Now, let's see...plants use sunlight to break down CO_2 and water to make sugar and oxygen. Where's the gasoline part of it?"

"Not *gasoline*, you cretin! Hydrogen!"

"What hydrogen?" I shot back. "It all goes into making sugar!"

"Not all of it," he said, drawing heavy black lines in the air with his finger. They persisted like anchored rods of artfully curved graphite, despite the steady breeze. "In fact, the standard chemical equation for photosynthesis looks like this:"

$$2H_2O + CO_2 + \text{sunlight} \rightarrow \text{sugar } (CH_2O) + O_2 + H_2O$$

I studied the equation, then asked, "So what's CH_2O? It's not sugar."

"No, it's a basic building block of sugar. If that's confusing, let's make some glucose and write it like this..." With a wave of his hand the old equation disappeared, and was replaced with:

$$12H_2O + 6CO_2 + \text{sunlight} \rightarrow C_6H_{12}O_6 + 6O_2 + 6H_2O$$

That made more sense, but it was still confusing. "Why is there water on both sides of the equation? Why not just take out half the water from the left side and erase it from the right side?"

His eyes narrowed, and he grinned smugly. "Because it's not the same water," he answered.

"Huh?" was my measured response.

He said, "Here; watch carefully..." His fingers went into a little dance, and quickly the $12H_2O$ on the left side of the equation became 12 of the familiar lopsided water molecules. In likewise fashion, the $6CO_2$ became six linear carbon dioxide molecules. From out of nowhere, a bright shaft of sunlight bathed the small molecules in intense, yellow light, and they all dissociated into free atoms. I counted 24 each of the hydrogen and oxygen atoms, and 6 carbons. Zed snapped his fingers and all the atoms reformed into molecules on the other side of the equation. Six of the 12 oxygen atoms from the CO_2 went into the glucose molecule, while the other six went into the water. To my surprise, all of the oxygen from the old H_2O was liberated as O_2.

"It's all *new* water, made using the oxygen in the CO_2, rather than what was in the original water. That means *all* the old water is broken down, so there's a point early on in the reaction where there's a lot of free hydrogen and oxygen, as you just saw. And it just so happens this particular algae releases some of it without turning it back into water. In other words, the end of the reaction would show free H_2 and O_2, and less water." He paused, then asked, "Is any of this getting through to you?"

"Yeah. This stuff exhales hydrogen gas...like you said."

He let out an exasperated sigh, and continued. "The problem is, the enzymes that liberate the hydrogen shut down when there's too much oxygen around. So the trick is to genetically engineer the enzymes to tolerate oxygen. Once that little kink gets worked out, there could be vast

Zed sez

There could be vast floating carpets of this stuff in tidal pools, converting sunlight and sea water into car fuel at 15 percent efficiency, or more.

floating carpets of this stuff in tidal pools, converting sunlight and sea water into car fuel at 15 percent efficiency, or more. That's better than the efficiency of a standard solar panel, without having to worry about the wasteful business of making electricity."

"Sorta gives you a new respect for pond scum," I replied, dipping a finger into the slime, and wondering how many gazillion hydrogen molecules were drifting past it, on their way to interstellar space.

"An added advantage of biological systems is that they are self-perpetuating and require no additional energy input, beyond what nature provides.

"On the downside, however, you're vulnerable to the ravages of nature. Ask any farmer what nature can do to a nice productive field. Violent storms can rip your algae bed to pieces. You also have to worry about contamination, disease, and invasion by opportunistic species. For those reasons, many people prefer to get their renewable energy from sources with more of a, shall we say, engineered element. That's why we're also working on photoelectrochemical water splitting."

"Photo...*what?*"

Photo...oh, I'll just show you. Follow me." I trailed behind him to the far side of the big tank, where a separator in the water prevented the algae from spilling over. "Here, what do you see?"

Looking down into the tank, I said, "It looks like a solar panel, dunked in a water trough."

"How observant of you. But this isn't just any solar panel, and that's not just any water."

I looked again. "It isn't?"

"No. The water is an electrolytic solution, and the solar panel is doped with a metal catalyst that works like an electrolyzer. The idea is to use the electricity produced by the solar panel to electrolyze water directly, without having to hook it to an expensive stand-alone electrolyzer."

I said, "I can see it all now: squads of energy police kicking all the houseboats and water skiers off Lake Mead, so they can submerge a million solar panels 3 inches below the surface. That'd raise a few short hairs, I'll tell you what!"

Zed rested his hands on his pinbones and said, "Don't be ridiculous! Everyone knows we're going to use the Great Salt Lake."

Astonished, I whispered, "Really?"

"Ha! Gotcha!" he barked, triumphantly. "Now shut up and listen, will you? It won't do to just submerge a bunch of solar panels in a lake. You need to set up vast ground-based arrays, like this one over here, except much larger." With a jaunty gait, he led me to the backside of a dense hedgerow of juniper bushes, where I saw several rows of odd-looking solar panels. Rather than being customarily flat, they all had rippled surfaces, as if they'd been fashioned from corrugated glass. They were all connected together by a series of pipes of different sizes. Many led to something that looked like a compressor. Beyond the compressor stood a stout, round tank. Next to that there was a golf cart with an unwieldy array of equipment mounted in the back.

"It takes specially designed solar cells with transparent channels to permit the passage of light and the flow of electrolytes, and membranes to separate the H_2 and O_2 gases. It's a real technological challenge. For a truly efficient unit, you need to use multilayered cells, each layer of which is activated by a different wavelength of light. At this stage in the technology, it requires some very exotic substances. But, once the bugs are worked out, it should be possible to mass produce devices that operate at 25 percent efficiency or better."

I did some quick calculations in my head. "Okay, let's assume that the solar panels that run my house operate at about half that—at 12.5 percent efficiency. And then figuring that the power lost to the components—the charge controller and inverter—knocks that down even further, to say, 10 percent. If, on average, I produce 4 kWh per day with 100 square feet of panels—and that's a low estimate—then with this photochemi-whatsit

Zed sez

A highly efficient fuel-cell car that gets the equivalent of 70 miles per gallon of gasoline would get only 20 miles on one gallon of liquid hydrogen. But, by the same token, six pounds of hydrogen—the weight of one gallon of gasoline—would propel that same car over 200 miles.

system I would produce the equivalent of...what?...10 kWh's per day of hydrogen? That's..."

"...About 34,100 Btu of hydrogen," Zed chimed in, "or roughly one gallon, if it were liquefied. Over an area of one square mile—leaving about half the area open so one row of arrays doesn't block the light falling on the array to the north of it—such a system would make about 140,000 gallons of hydrogen per day. That's where you live, in Colorado. Down south, in the Nevada desert? It could be over 180,000 gallons per square mile."

"That's a nice haul, Zed."

"Indeed it is. If you estimate that an efficient fuel-cell vehicle could get 20 miles per gallon of hydrogen..."

"Hold on a minute!" I objected. "What's this '20 miles per gallon' business? I thought hydrogen was an efficient fuel!"

"And it is, my hot-headed young friend. Pound for pound, hydrogen contains nearly 3 times the energy as gasoline. But a gallon of liquid hydrogen only weighs about 0.60 pounds, compared to six pounds for a gallon of gasoline, so it takes nearly 3.5 gallons of hydrogen to equal one gallon of the smelly stuff you currently put in your pickup. Comprende? A highly efficient fuel-cell car that gets the equivalent of 70 miles per gallon of gasoline would get only 20 miles on one gallon of liquid hydrogen. But, by the same token, six pounds of hydrogen—the weight of one gallon of

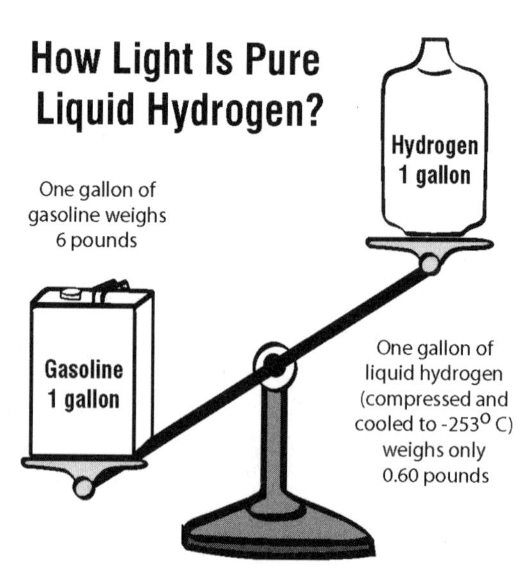

gasoline—would propel that same car over 200 miles."

"Sure. I knew that," I said, feeling like a red-faced fool. And a liar.

He glanced at me with a furrowed brow, then said, "If we assume that most cars travel somewhere in the neighborhood of 32 miles per day, and then figure about a 10 percent loss in transportation and distribution—a very pessimistic figure—then one gallon of hydrogen would last our efficient fuel-cell car for 0.56 days. So 180,000 gallons of hydrogen per day—the yield from our one square mile of super-efficient photoelectrochemical cells—would propel nearly 100,000 cars...forever."

"I think we're making some progress, Zed."

"Good. When it comes to hydrogen, the word 'progress' loses its taint. Now, let's review what you've learned so far about the ways of liberating hydrogen from water."

"Okay," I chimed in, "you can split water with electricity..."

Zed nodded. "And there are endless variations on that theme. You can enhance the process with chemicals, heat and pressure...in more ways than you can imagine."

I added, "And with the right solar apparatus, you can even bypass the need for an electrolyzer."

"Correct. Then there's the biological pathway. By growing the right organisms under the right conditions, Mother Nature will do the work at ambient temperature and pressure, without the need of electricity, like she's done for more years than you or I can count."

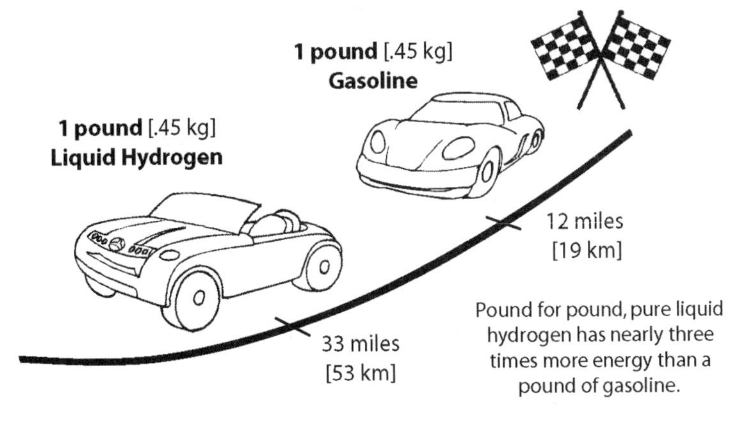

Comparing Weight of Hydrogen and Gasoline

"So that about covers it, huh?" I surmised, thinking there couldn't be much more you could do to split water.

"Not quite," Zed said, with a coy grin that always meant something unexpected was lurking around the corner. "C'mon. There's something I want to show you."

I followed Zed to the end of the solar field, to the waiting golf cart. "Let's take a quiet little ride," he said, sliding into the driver's seat. Why not? I sat down beside him.

No sooner had I situated myself than Zed put the pedal to the metal, and the cart's instant acceleration slammed me back into my seat. It zipped along like a go-cart with a pair of beefy—but somehow silent—chainsaw engines, and Zed drove it accordingly. There being no road to follow, he ran a white-knuckled slalom course around rocks and bushes with the intensity and aplomb of a downhill skier. I grabbed hold of whatever I could, thinking that when the inevitable occurred I could at least propel myself to

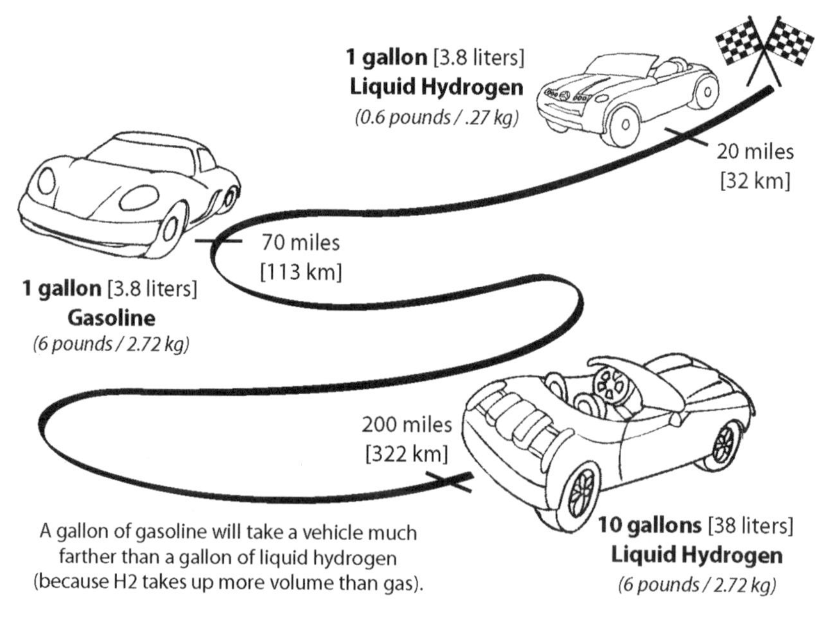

A gallon of gasoline will take a vehicle much farther than a gallon of liquid hydrogen (because H2 takes up more volume than gas).

Comparing Volume of Hydrogen and Gasoline

NOTE: There are many ways under consideration for storing hydrogen. While it is convenient to compare liquid hydrogen to gasoline, this form of hydrogen has drawbacks. See Chapter 15 for a detailed discussion of other options for storage and transportation.

where I wanted to land, without a turbo-charged, hydrogen fuel-cell powered golf cart on top of me. Zed, of course, enjoyed every second of it.

I was hoping he'd head into the interior of the Wasserstoff Farm, where I might get a better glimpse of what lay at the heart of this strange complex, but instead he hugged the perimeter until, after a drive of less than a mile, our destination came into view. At first I thought it was some sort of New Age outdoor chapel, since from a distance it had the appearance of rows of curved seats laid out in a semi-circular pattern in front of a brightly-lit altar. But as we drew closer I saw that the seats were all curved mirrors which directed sunlight toward a single mirror mounted on top of a narrow tower. The light from the tower-mounted mirror was focused downward to a glowing, cylindrical vessel mounted near the ground. The light reflected from this vessel was so intense it was hard to look at.

Zed slowed the golf cart to a stop and we both got out. "Don't touch anything, and be careful where you direct your eyes. It's all very hot and bright," he announced. "*This* is direct solar thermal water splitting."

"Is that even possible?" I asked, finding it hard to believe such a thing could be done.

"Oh, yes. Given enough heat, anything breaks down. Water begins to dissociate at a little under 1,700 degrees Celsius. The idea is to concentrate the sun's energy several thousand-fold, until the temperature inside the reactor vessel can be sustained at about 2,230 degrees Celsius. At that temperature about 25 percent of the steam will be split into hydrogen and oxygen."

"Why not all of it?" I asked, puzzled.

Zed answered, "Heat imparts energy to a system. In a general sense it will make all of the molecules more energetic, which means they'll move faster and be more apt to run into each other. But what we refer to as temperature is merely an indicator of the *average* energy of a system. Within the system—in this case, the reactor vessel—the individual water molecules will have a wide spectrum of energies. Only those with the highest energies will split apart."

It made sense.

"Of course, 2,230 degrees Celsius is exceedingly hot—700 degrees hotter than the melting point of iron, in fact. So you don't just make a reactor from off-the-shelf hardware items. A quartz jar lined with zirconia—zirconium dioxide, or ZrO_2—is what's needed. Then you have to

separate the gases from one another and move them away from the reactor without taking too much heat along with it. It's a simple idea that requires a complex implementation. But it does have its advantages."

"Such as?" I asked.

"Well, it makes good economic sense to concentrate the sun's energy with reasonably inexpensive mirrors, rather than making large, costly surfaces to react with unconcentrated sunlight. The downside is that clouds can pretty much shut down the whole operation, so it's only economically feasible in hot, arid regions. Fortunately, there's no dearth of such places in the southwestern United States. It's not the basket I'd want to put all my eggs in, but it's certainly another piece to the puzzle."

Eggs and puzzles? "Aren't we mixing our metaphors, here, Zed?" I laughed, hoping to see a sheepish look wash across his face.

It never happened. Instead, he said, "Just trying to keep you on your toes, lad. Glad to see your synaptic links are still intact."

Yeah, right.

technistoff - 9

Car fuel from pond scum? When Zed first suggested it, I thought perhaps he'd made too many trips through the Dimensional Compressor. But then I got to thinking, *why not?* It doesn't really take too great a leap of faith, once we stop to consider that almost all of the fuel we use for power production today is the end-product of massive ancient reservoirs of rotting vegetation. Or that the greater part of what we consider food begins as water and atmospheric carbon dioxide. It's all part of the chemically mystical dance of hydrogen, carbon and oxygen, as they are forever called by sunlight from their lethargic, bonded states to do some biological deed, before being allowed to fall, yet again, back into their low-energy slumbers.

biological: water splitting by microorganisms

The part that's hard to believe is that we, accustomed as we are to taking 3,000 pounds of metal with us everywhere we go, could derive *enough* energy from these quirky microscopic organisms to do us any good. But the numbers—theoretical as they are at this point—suggest that it just may be so.

As a matter of clarification, the chemical equation Zed drew for me merely shows the beginning and end products of a long, complex series of reactions, involving dozens of intermediary compounds, known collectively as photosynthesis:

Photosynthesis: $12H_2O + 6CO_2 + \text{sunlight} \longrightarrow C_6H_{12}O_6 + 6O_2 + 6H_2O$

There are a number of researchers exploring numerous ways to tap into the abilities of certain organisms to liberate hydrogen from water. The process that most closely resembles the one Zed explained was described in *Molecular Engineering of Algal H_2 Production*, a paper presented at the Proceedings of the 2002 U.S. DOE Hydrogen Program Review.

As Zed suggested, the ultimate success or failure of this specific line of research lies with the researchers' success in reengineering a pair of H_2-producing enzymes to be more tolerant of O_2. Their goal is to reach 20 percent efficiency (a bit more optimistic than the 15 percent Zed suggested) by the year 2010. They've got a long road ahead of them. To read the paper noted above, or simply for more information about this, and similar projects, visit the National Renewable Energy Laboratory (NREL) website at: *www.nrel.gov/hydrogen/proj_production_delivery.html*.

photoelectrochemical

Nor are we likely to see mass-produced solar arrays, capable of converting sunlight to hydrogen with 25 percent efficiency anytime soon. But who knows? There's certainly no lack of effort out there by people trying to make it happen. And, if scientific breakthroughs can be likened to earthquakes, then solar-cell research is overdue for The Big One.

A fairly rigorous paper on this subject, titled *Photoelectrochemical and Organic-based Solar Cells* can be found at: *www.unict.it/dipchi/05Didattica/Corsionline/Coloranti/13_Fotochim/solarconv/panela1.htm*.

For more digestible discussions on this line of research, visit the NREL website listed above, or: *http://gcep.stanford.edu/pdfs/hydrogen_workshop/MacQueen.pdf*.

The figure of 4 kilowatt hours per day per 100 square feet *[9.3 square meters]* of surface area is a tidbit of knowledge I plucked from my own experiences with solar energy. The actual figure for my 1,140-watt array is 4.2 kWh/day, averaged over the course of a year.

Zed was mum on the subject of his fuel-cell-powered golf cart, so I can't tell you where to get one. Sorry.

direct solar thermal water splitting

And finally, is it really possible to create temperatures in excess of 2,000 degrees Celsius with an array of heliostats (curved mirrors)? Anyone who has ever played with a cheap magnifying glass and a pile of dried grass should be able to answer that one. Perhaps the most promising work in the field of direct solar thermal water splitting is being done in Israel at the Weizmann Institute of Science (*http://80.70.129.78/site/EN/homepage.asp*), as described in the November 1996 *Hydrogen & Fuel Cell Letter* (*http://hfcletter.com/archives/nov_96.htm*).

The direct solar thermal approach is not without its drawbacks, of course. Achieving reactor temperatures of over 2,000 degrees Celsius is not easy. Nor is it a simple matter to separate the hydrogen from the oxygen, once the reaction has taken place. A promising variation on this theme uses concentrated sunlight at 1,750 degrees Celsius to reduce zinc oxide to metallic zinc and oxygen. Then, in a separate, non-solar reaction, the zinc is reacted with water to make pure hydrogen and…what else?…zinc oxide, which can then be recycled. By using carbonaceous materials (such as biomass) as reducing agents, the reaction can be made to proceed in the relatively low temperature range of 1,000 to 1,400 degrees Celsius. The metallic zinc does not have to be reacted with water immediately, of course. It can be transported to where the hydrogen is needed, or even used to make electricity directly in a zinc-air fuel cell.

For a visual description of this process, visit the ASES website at: *www.ases.org/hydrogen_forum03/Steinfeld.pdf.* For a more explanatory paper on the subject, read *Solar Carbothermic Production of Zinc and Power Production Via a ZnO-Zn-Cyclic Process,* at: *www.solarpaces.org/ISES_O6_14_Solzinc.pdf.*

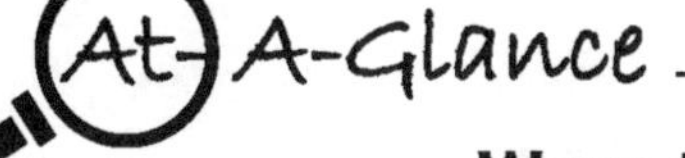

Ways to Liberate Hydrogen

- ★ **Electrolysis** - split water with electricity
- ★ **Steam Electrolysis** - split water with heat, pressure and electricity
- ★ **Direct Solar Thermal Water Splitting** - split water with heat
- ★ **Photoelectrochemical** - split water using sunlight directly
- ★ **Thermochemical** - split water using chemicals and heat
- ★ **Biological** - split water using microorganisms
- ★ **Steam Reforming** - convert methane in natural gas using steam
- ★ **Direct Thermal Splitting of Natural Gas** - split natural gas using heat
- ★ **Gasification** - breakdown coal or biomass with heat and pressure

Here on the ground we're still getting from place to place using 19th century technology—internal combustion engines—and fuel that was originally a by-product of kerosene manufacture. NASA, on the other hand, has been using hydrogen to power their rockets and provide on-board electricity for decades.

chapter 10

(Re)forming a Steamy Relationship with Hydrocarbons

I watched with interest as Zed prepared what I hoped would be a simple demonstration. After the wild ride back over rocks and through the bushes in Zed's Formula-One golf cart, "simple" had taken on a whole new appeal. "Stationary" was also a word I was quickly warming up to. Yessiree; nothing quite as nice as having two feet on terra firma.

Still, from the looks of it, this lesson was going to be a little bit too simple. It was the classic example every student sees in every textbook that ever explained the concept of energy. Put a rock at the top of a hill, then try to impress your students by telling them the rock is imbued with some mysterious quality called "potential energy." Give it a little shove, and down it goes, to the bottom of the hill. Now it's used up all that potential energy and, unless there's yet another hill to roll it down, the rock is pretty much worthless in the energy department.

Ho hum.

"Okay. Here we go," Zed announced. He hadn't bothered with hills, I saw. Instead, there were eight variously-sized rocks resting on the same level, in two groups about one foot apart. In the first group, one large rock and two small ones rested on the edge of a 3-foot drop-off; in the second group, one large rock and four small ones were poised by a drop-off of only a few inches. Zed said, "You can think of the first group of rocks as the two hydrogen atoms and the single oxygen atom in water."

"Sure, whatever," I yawned.

His disgust at my indifference was obvious as he said, "Fine! You don't like rocks...?" With a wave of his hand, the rocks rose from their resting places and glided into his hands. He promptly dropped them on my feet.

"Ow! That hurt!" I cried.

"Quit whimpering. It's only *potential* pain," he said, with a nasty smirk.

"Feels real to me," I moaned.

"Potential *is* real. That's the point."

"You went to all that trouble to drop rocks on my feet?"

"No. I dropped rocks on your feet to get your attention. It seems to have worked."

"Boy, howdy," I groaned.

"So, let's start again," he said, eyeing me to see if I was paying attention. I was. "I've replaced the rocks with molecules. Happy? Here, where we formerly had the two little rocks and one big one, we've now got a swarming hive of white H_2 and blue O_2 molecules. Is that more to your liking?"

"Uh, yeah," I said, thinking this might get interesting yet.

"Good. Now, over here we have carbon and hydrogen." The hydrogen was again in the form of H_2, but the carbon atoms, large and red in color, were unbound—a most unnatural state for carbon to be in. "Both groups of elements have roughly the same potential energy at this point, but there's one significant difference between the two."

"I'm listening," I told him.

"I know you are, lad," he said, lighting a match with his thumbnail. "Now watch..." With his free hand he scooped all the H_2 and a handful of O_2 molecules out over the cliff, where they hovered in midair. Then he lit the match. Instantly the hydrogen and oxygen molecules united with a flash and a *whumph*, and fell 3 feet to the ground in a pool of water. "So tell me: what just happened?"

"Well, you introduced energy into a volatile system, causing a reaction

Zed sez

Since the bonds of methane, and, by extension all hydrocarbons, are so easily broken, it takes a lot less energy to free up the hydrogen in those molecules. That's why most commercial hydrogen in the world today is produced by breaking down hydrocarbons.

that forced the system to give up its potential energy and collapse into a very low energy state."

"Very good!" he said approvingly. "And what would you have to do to get useful energy back out this system?"

I eyed the puddle of water, and said, "You'd have to add at least as much energy to the pudd...uh, system, as you just took out of it. It's the only way you're going to free up the hydrogen and oxygen and get them back on top of the cliff."

Zed was pleased. "Exactly!" he exclaimed, "I should drop rocks on your toes more often."

"I'd rather you didn't," I opined.

"Now, to the hydrogen and carbon..." He took his hands and curved them around the swirling atoms, squeezing them together like a handful of bees. In a moment he had made a batch of new molecules, each with four hydrogen atoms evenly spaced around a central carbon atom. They slid off the cliff to the ledge a few inches below. "This is methane, the precursor to all hydrocarbon molecules. Can you see how it differs from water?"

"Well, aside from its appearance, it still has a lot of the potential energy of its constituent atoms," I said.

"Right. Unlike water, you can still get energy out of it in a sustained reaction. The bonds are higher energy bonds than those of water, so it's easier to break them."

"Understood," I agreed, "but what's the point?"

"The point, my young friend, is exactly this: since the bonds of methane, and, by extension all hydrocarbons, are so easily broken, it takes a lot less energy to free up the hydrogen in those molecules. That's why most commercial hydrogen in the world today is produced by breaking down hydrocarbons. Compared to water," he said, looking at the puddle on the ground, "there's more energy in the hydrocarbon molecule to begin with, so it's faster and cheaper to break it down."

"Yeah...all you have to do is light a match," I said.

"You can, but it won't liberate any hydrogen. At least, not for long."

"Why not?" I wondered, "there's nothing there but hydrogen and carbon. What else can it do?"

In answer to my query, Zed struck another match and lit the airy pile of methane molecules. With a *ka-whumph* familiar to anyone who has ever lit a gas stove, the methane dissociated into water and carbon dioxide and rolled off the ledge, down to the lowest energy level.

"Hold on!" I protested, "where'd all the oxygen come from?"

"The air, my budding genius friend. In case you've forgotten, the atmosphere is full of it."

Oh, yeah.

Zed continued. "So the trick, then, is get the hydrogen from the hydrocarbons back up to here..." he motioned to the top of the cliff, "...without letting it slip down to here." He nodded to the puddle of H_2O and CO_2 at my feet.

"How?" I asked.

He answered, "Obviously, not with fire, at least not when there's any oxygen about."

"Obviously," I echoed.

"So the next best way to add energy to a population of gas molecules is with steam. In fact, most of the hydrogen produced in the United States is made by a process called 'steam reforming of natural gas'."

I held up my hand, and said, "Excuse me, Zed, but why are we even talking about this? I thought the whole idea of the 'hydrogen economy' was to get away from fossil fuels. Make hydrogen from water with solar and wind power, and all that. Making it from natural gas seems to me to be, well...counterproductive."

He scratched his head, then said, "You need to reexamine your premise, lad. Is the idea to get away from fossil fuels, or is it to get away from the problems caused by fossil fuels? There's a difference. Nor is there a single goal; nothing is ever that simple. The ultimate goal may indeed be to leave all the world's remaining fossil fuels safely buried in the ground. But the immediate goal is—or should be, at any rate—to reduce greenhouse gas emissions as quickly as possible."

I opened my mouth to speak, but Zed snared me with narrowed eyes, saying, "Before you say anything, let me remind you that there's a field of ice poised over the gulf stream of sufficient volume to make a respectably-sized moon. And it's getting warmer. It would be a real pity if Europe were cast back into an ice age while everyone's driving around in gasoline-powered automobiles, wondering how to produce the requisite amount of hydrogen from renewable sources."

I could see his point. I asked, "Is it really *that much* easier to make hydrogen from fossil fuels?"

"Wrong question," he answered. "Easy has nothing to do with it. The electrolysis of water is the epitome of easy. But it's also wasteful. First you have to produce the electricity, which in this day and age usually means burning coal to produce the heat to create the steam that turns the generator that cranks out the kilowatts. You also have to consider that water is in a lower energy state than methane to begin with, so it takes more energy to liberate the hydrogen it contains. And when you're talking about energy on the scale of the 26 quadrillion Btu being used by the transportation sector in this country...well, you figure it out. If it's really true that an efficient hydrogen fuel-cell vehicle can expect to get the equivalent of 70 miles per gallon of gasoline, then the sooner we get those kinds of vehicles on the road and get this hydrogen thing rolling, the better."

He crossed his arms, and said in conclusion, "A less-than-perfect initial effort now is far superior to waiting until everything *is* perfect. Don't you agree?"

Put in those terms, it was hard not to. "Absolutely," I said.

"Good. Now we can get on with the chemistry of the matter. Conceptually, the steam reforming of natural gas is a simple process. All you need is methane, steam—heated to about 850 degrees Celsius, about 100 degrees below the melting point of silver—and pressure."

"How much pressure?" I asked.

"About 2.5 megapascals."

"Can we speak English, please?"

"Oh, alright. One megapascal is about 10 atmospheres, or 145 pounds per square inch. So two and a half of them would be 25 atmospheres, or around 360 psi. Ten times the pressure in your pickup tires. Happy?"

"Elated."

"Anyway," he went on, "with all that heat and pressure, you have gazil-

Zed sez

A less-than-perfect initial effort now is far superior to waiting until everything is perfect.

lions of very energetic molecules crammed very tightly together. It's a reaction waiting to happen."

"And *then?*" The suspense was killing me.

"And then you pass this stream of hot, crowded molecules of methane—CH_4—and steam over a catalyst—usually nickel—and, *presto, chango!*, you have H_2, CO, and CO_2."

"CO? Why is there carbon monoxide?"

"Because, as chaos theorists are quick to point out, that's just the way the cookie crumbles. It's a reaction that proceeds rather quickly, you see, and there's just not enough oxygen to go around. You can lower the temperature and avoid a lot of CO production, but it's inefficient. Raising the temperature will increase efficiency but it will give you more CO. You have to strike a balance. But don't worry about the CO; we're going to get rid of it."

"How?"

"By reacting it with water and different metal catalysts, over two separate ranges of heat and pressure. It's called a 'water gas shift reaction', and it's exothermic, meaning heat is produced when the H_2O and CO are converted into CO_2 and H_2. Ideally, this heat can be captured and used in the reforming process."

"Where does the heat come from?" I asked.

"From the carbon monoxide. Believe it or not, it's somewhat combustible, though its Btu content is only about 7 percent of hydrogen. But the energy in CO does help the whole process along."

"It all sounds very complicated," I said.

"Oh, it is," he agreed, "and we're not done yet. Our hydrogen still isn't clean enough to use in a fuel cell, since any residual carbon monoxide will foul the very expensive platinum anode. So we have to put the mixture through one more step. This one is called 'pressure swing adsorption'."

I was getting confused. "Where ever do these names come from?"

Zed answered, "It's just the way of science; not a lot of room for humor, I'm afraid. But pressure swing adsorption really is an apt term. That's because the remaining impurities are collected in adsorbent columns under high pressure, and flushed out under low pressure. The pressure swings from high to low. Get it?

"Anyway, I wasn't intending to teach you how to make your own hydrogen from natural gas; I just wanted to give you an appreciation of the

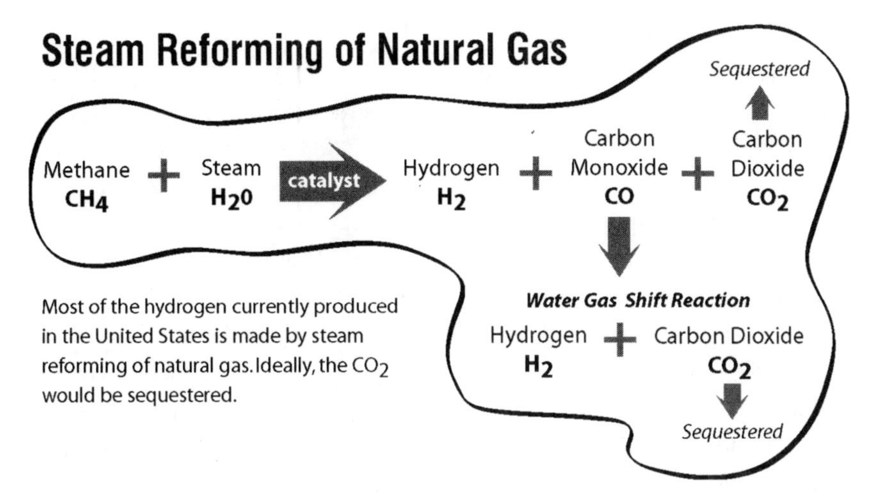

Most of the hydrogen currently produced in the United States is made by steam reforming of natural gas. Ideally, the CO_2 would be sequestered.

process. This isn't the only way to do it, of course, just the most common. One half of all hydrogen produced in the world is extracted from natural gas, by this and similar processes, and 95 percent of what's produced in the United States."

"What is hydrogen used for, if there aren't any fuel-cell cars to put it in?" I asked.

"Fertilizer," he answered. "Ammonia, NH_3, which is then turned into ammonium nitrate, NH_4NO_3, a solid soluble in water. The hydrogen keeps the nitrogen atoms from pairing up in their customary triple-bonded low-energy state. By joining with hydrogen and oxygen into a nitrate, the nitrogen is in a form plants can use.

"And, of course," he added, "hydrogen also makes a great rocket fuel."

"Sure, okay," I said absently, my mind now moving in another direction. Throughout Zed's description of the steam reforming process, my mind conjured up images of trillions of little CO_2 molecules, floating into the air and capturing heat reflected from the Earth. I asked, "What happens to all the CO_2? Is it just released into the atmosphere?"

"Depends," he answered. "It's easy enough to capture it, but then you have to figure out what to do with it. Sequestration underground in depleted gas wells is one possibility."

"A lot of people won't like that idea," I suggested.

"What's not to like?" he replied, with a questioning stare. "That's where the carbon came from in the first place."

"I don't know, Zed..."

With a wave of his hand, he dismissed the issue. "Whatever. Natural gas is not the future, anyway. It's getting hard just to find enough of it to heat our homes. And once we start using it for hydrogen production...? Besides, many would argue that by the time you go to all the trouble to break down natural gas and sequester away the CO_2, you'd have been better off just using the CH_4 as it was. As far as hydrocarbons go, it's about as clean a fuel as you can get. "

I was puzzled. "Okay, then. What else is there, besides natural gas?"

"Well, there's lots of things. As I'm sure you'll remember, anything that was once alive is going to be made up mostly of carbon and hydrogen. That includes all fossil fuels, including gasoline and oil, as well as biomass—everything from garbage to plastics, to forest and agricultural refuse—and then, of course, there's coal."

"*Coal?* How do you get hydrogen from coal? I thought it was nothing but carbon!"

"Oh, it's got hydrogen—about half the atoms in coal are hydrogen, in fact. But weight-wise it only works out to about 5 percent, compared to 25 percent by weight for methane."

"Big deal," I scoffed, "what good does 5 percent do you?"

"Obviously, you want to add more hydrogen to the process, so you pulverize the coal, mix it with steam, then react it under high temperatures and pressures."

"Fine, but you still haven't explained where the hydrogen comes from."

He answered, "Sure I have, you're just not listening. It comes from the steam."

I looked at the puddle of water on the ground and recalled Zed's little demonstration earlier. Especially the last part where the methane reacted with atmospheric oxygen and created water and carbon dioxide. And how counterproductive it was. "I thought water was a no-no," I said.

Zed rolled his eyes. "I said you didn't want to *make* water; I didn't say anything about *breaking* it."

"Oh...yeah."

"Alright, then. Now that we're clear, I'll continue." His eyes darted in my direction, to see if my mouth was going to open. It didn't. "Once the coal and steam have been gasified, you have a mixture called synthetic gas, or simply syngas. Syngas is made up of CO and H_2."

"That sounds familiar," I interjected.

"Indeed it does. So perhaps you can tell me what comes next?"

I answered, "If I were a betting man, I'd say we have a water gas shift reaction in our future."

Zed beamed with a professor's pride. "And you'd be right, lad! A reaction that, as you no doubt recall, makes CO_2 and more H_2 from CO and steam. When it's all said and done, you've made about 13 pounds of hydrogen for every 100 pounds of coal. That means 8 pounds of H_2 was produced from steam and 5 pounds from the coal, itself."

"What about sulfur and nitrogen compounds? Acid rain and all that? Isn't there a lot of that in coal?"

"Certainly enough to be troublesome. But it can be recovered."

I asked, "And what about all the CO_2? Did it slip away?"

Zed said, "Nah, we captured it all and dumped it in same hole as before."

"Hmmm..."

Zed cupped his chin between his thumb and index finger and studied me with his deep blue eyes. Finally, he said, "You've got a real thing about CO_2, don't you? Well what if I told you there was a way to make hydrogen that could actually decrease the amount of atmospheric CO_2?"

That caught my attention. "Decrease?" I echoed.

"Just so," he said.

"Biomass?"

"Exactly. We can use pretty much the same process we did for the gasification of coal. We just have to add a preliminary step, known as thermal

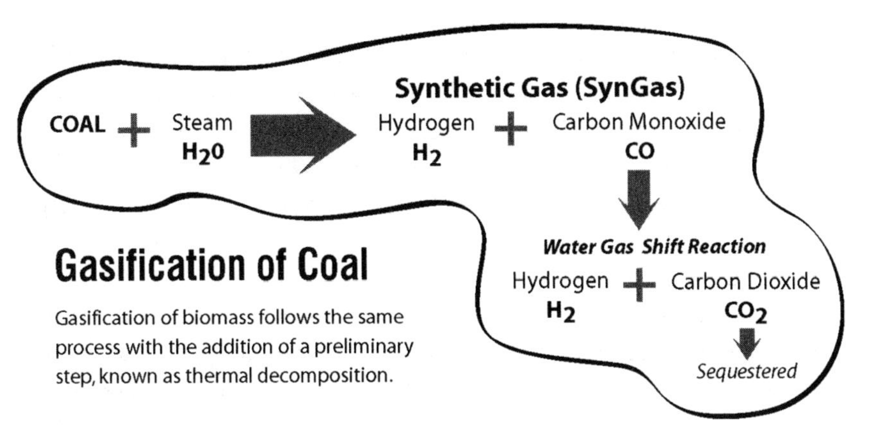

Gasification of Coal

Gasification of biomass follows the same process with the addition of a preliminary step, known as thermal decomposition.

decomposition. This is where the organic matter is converted into gases, condensates and coke. As I said before, you can use about anything, but in order for the process to be productive, it makes sense to use things that are in abundance, such as forest and agricultural residues, from farming and forest thinning operations—wood chips and corn stalks, in other words. Another possibility is to grow fast-growing, drought-resistant crops on fallow or marginal land."

"Sounds good, but what's this got to do with decreasing atmospheric CO_2?" I asked.

Zed replied, "Think about it...you burn gasoline in your pickup. The carbon you're burning was buried in the ground eons ago, but you've dug it up, mated it with oxygen, and set it loose in the atmosphere. How do you capture it, and put it back into the ground where it belongs?"

"Probably not with a butterfly net."

"Uh, no. You have to commission a green plant to do it. Then you harvest the plant and gasify it, extracting the hydrogen it contains. In the process you sequester all that carbon and..."

"Stick it where the sun don't shine!" I proudly exclaimed.

"Yeah, well, that's one way to do it," he said.

"There's another way?" I asked, feeling kind of flustered, like maybe Zed had been holding out on me.

"Sure. You can turn it into fertilizer."

"Say *what?*"

"Fertilizer. It's real simple. During the thermal decomposition—or maybe we should say, the pyrolysis—step, we take most of the carbon from the plant tissue and bind it up into charcoal pellets..."

"Charcoal?" I scoffed, "what good is *that?*"

He speared me with a cutting glance, and replied, "More good than you can imagine, my skeptical apprentice. Charcoal traps nutrients added to soil, and significantly reduces the leaching of nutrients already there. It also makes a great supporting framework for microbial growth, such as the bacteria that fix atmospheric nitrogen. And, to top it off, charcoal increases the soil's capacity to hold water."

In my defense, I said, "That's great, Zed, but that's really not what fertilizer is all about. I mean, give me some good old ammonium nitrate and I'll show you a thing or..."

Abruptly, he cut me off. "I was just getting to that."

"Oh." I squeaked.

"You see, you take the nitrogen liberated in the process of making syngas and use a portion of your hydrogen to make NH_3, ammonia." He paused to see if I was sufficiently attentive. Apparently I passed the test. "The ammonia gas is saturated with water vapor to hydrate it, then mixed with the charcoal pellets. As you might imagine, the charcoal soaks up the ammonia like a bureaucracy soaks up taxes."

"But what about the CO_2 left over from the process of making syngas? What happens to that?" I wondered.

Zed replied, "I'll tell you, just as soon as you shut up and let me finish. You see, this is the truly beautiful part; for once the charcoal pellets are saturated with ammonia, they're flooded with CO_2 gas and—presto-chango!—your charcoal is now permeated through and through with ammonium bicarbonate, NH_4HCO_3, a relatively harmless chemical used for a range of things from fire extinguishers to leavening for bread. But bound up in charcoal and returned to the soil, it becomes fertilizer. And, as icing on the cake, you've returned all the original trace minerals to the soil."

I thought for a moment about the formula, NH_4HCO_3, and said, "You know, Zed, NH_4HCO_3 can be rearranged into NH_3, H_2O, and CO_2; ammonia, water and carbon dioxide. You're making a fertilizer that gives off CO_2 as it's used up."

Zed threw his hands in the air, and retorted, "Who cares? You've got 80 percent of your carbon bound up in the charcoal, and it takes thousands of years for *that* to break down, so you're still taking four times as much carbon out of circulation as you're putting back in! And, once you factor in the other benefits of the charcoal, you're increasing plant growth by such a degree that most of the extra CO_2 released in the breakdown of ammonium bicarbonate is taken up again in a larger, more luxuriant plant."

"Ingenious, Zed. Really cool." It was, I had to admit.

He slapped me on the shoulder, and said, "And so it is, lad. I'm just glad we could get you past the notion that all sequestered carbon has to be buried deep underground."

"*Me?* Zed! It was..." From the smug smirk on his face, I could see it was hopeless to argue.

I didn't even try.

technistoff - 10

I have a confession to make. Though it may appear that the greater part of Zed's lesson about deriving hydrogen from fossil fuels and biomass was conducted in an uncharacteristically straightforward manner, it was anything but. As usual, Zed kept me visually rapt throughout the afternoon as he unraveled the intricacies of the various processes he described. My decision to overlook the details of his colorful wizardry was difficult but necessary, for were I to have given a full accounting of Zed's lesson, the essence of steam reforming of natural gas and the gasification of coal and biomass would have been buried beneath a mountain of visual magic. I hope you will forgive me.

For those of you who would protest that Zed's demonstration of potential energy was misleading, in that the relative heights between water, methane, and the free atoms of hydrogen, carbon and oxygen was inaccurate, I would have to concede that technically, you are right. Pound for pound, hydrogen has over two and half times the caloric (heat) value as methane (61,000 Btu/lb. versus 23,800). This is because 75 percent of methane's bulk can be attributed to the single atom of carbon, which has a relatively low caloric value (14,100 Btu/lb.). To be perfectly accurate, H_2 should have been on the highest level, carbon and H_2 together on a lower level, and the methane formed from them more than halfway to the ground. Oxygen should have been assumed (or not assumed) for both reactions. But really, isn't this getting confusing? Zed merely wanted to make the point that the hydrogen in methane is more easily liberated because methane is naturally in a much higher energy state than water. And so he did.

steam reforming of natural gas

Steam reforming of natural gas, and the subsequent processes Zed described for cleaning the hydrogen thus liberated, are tried and true technologies which, in Zed's judgment, served well to illustrate the fundamental concepts behind CH_4 to H_2 reactions. But steam reforming is not the only method out there.

There is an intriguing and potentially very promising technology called the Kvaerner Carbon Black and Hydrogen Process. It was developed throughout the 1980s and 1990s by Aker Kvaerner ASA, a Norwegian-based company. The

Kvaerner Carbon Black and Hydrogen Process

Kvaerner process sidesteps the messy question of "what do we do with endless tons of CO_2?" by simply eliminating it from the equation. Using a plasma-arc torch operating at 1,600 degrees Celsius to break methane and other light hydrocarbons directly into hydrogen and carbon black, the process produces no emissions, other than those resulting from the generation of the electricity used to run the torch (an amount equal to 1.25 kWh per cubic meter of H_2). Why doesn't the hydrocarbon feedstock simply oxidize into water, CO and CO_2? By using an electrically-driven plasma rather than a standard flame, oxygen can be excluded from the process.

The carbon black produced (250 kg of carbon black is produced for each 1,000 cubic meters of hydrogen gas) can be readily sold for use in rubber and tire manufacturing, and the excess steam produced is likewise a valuable commodity. Information about the Kvaerner Carbon Black and Hydrogen Process can be found at: *www.bellona.no/imaker?id=11196-1*, *www.akerkvaerner.com* and *www.hyweb.de/Knoewledge/w-i-energiew-eng3.html.*

solar-thermal dissociation of natural gas

For those of you who prefer to use sunlight to dissociate your natural gas, read the 2002 DOE/NREL paper by Jaimee Dahl (and many others) titled, *Rapid Solar-thermal Dissociation of Natural Gas in an Aerosol Flow Reactor*. It can be downloaded at: *www.eere.energy.gov/hydrogenandfuelcells/pdfs/32405a15.pdf.*

In this scheme the researchers concentrate 10,000 watts of solar energy on a solar reactor consisting of an outer quartz tube, and an inner graphite tube through which the natural gas flows. Argon gas flowing between the quartz and graphite tubes prevents oxidation of the natural gas, which is heated to temperatures in excess of 1,700 degrees Celsius.

As with the Kvaerner Process, the only products of the reaction are hydrogen gas and carbon black. The researchers suggest that 1,650 aerosol flow reactor plants, each producing 1,670,000 kilograms of hydrogen per year, could provide enough carbon black to satisfy the world's demand, thereby avoiding the consumption of 760 billion megajoules of fossil fuel per year (around 720 trillion Btu, or the equivalent of 5.76 billion gallons of gasoline), now being used in the production of carbon black. The dumping of 38 billion kilograms of CO_2 into the atmosphere would also be avoided. The hydrogen produced (1.67 million kg x 1,650) would provide 37 trillion Btu of energy. How many cars would that run? Well, if we assume an average car travels 11,000 miles per year, and that a super-efficient fuel-cell car gets the equivalent of 70 mpg of gasoline (with an energy value of 125,000 Btu per gallon), then each car would require about 19.65 million Btu per year. By doing the math, we see it would take over 1.88 million cars to burn up 37 trillion Btu of hydrogen. Not bad.

Of course, it'll take more than a couple of truckloads of natural gas to make it all work.

coal gasification

Coal gasification technology was originally developed as a cleaner and more efficient means of producing electricity; using it for the production of hydrogen has always been a possibility that was built into the process, but something of an afterthought.

Most coal-fired power plants today burn coal, crushed into a fine powder, in a large combustion chamber. The heat of combustion is used to create large amounts of high-pressure steam, which is channeled into a turbine, where it expands and turns the turbine's blades. Electricity is produced by the electrical generator attached to the turbine, while the steam—which has lost much of its energy by this point—is recirculated back through the boiler, where it picks up a fresh supply of Btus from the burning coal.

All in all, it's a rather old and crude technology—no different in principle than a 19th century steam locomotive—and something of an embarrassment for any society that prides itself on its technological prowess. It's also highly polluting, since every pound of coal that's burned releases nearly two pounds of CO_2 into the atmosphere, along with a slew of sulfur and nitrogen containing compounds.

According to James E. Hansen of the NASA Goddard Institute for Space Studies, "Coal is both the principle root of the CO_2 climate problem and the potential solution." In *Can We Defuse the Global Warming Time Bomb?* Hansen argues that there is 10 times more potential CO_2 in the world's coal resources than in its oil resources, and if we could just avoid the emissions from coal we could successfully derail runaway global warming. (Hansen's article is available for download at: *http://pubs.giss.nasa.gov/docs/2003/2003_Hansen.pdf*. See also Technistoff 6.)

syngas

This is where coal gasification comes in. Coal gasification plants are more efficient, and potentially far cleaner. The hydrogen created in the gasification process can be separated from the carbon monoxide in the syngas (a mixture, you will recall, of CO and H_2) and burned in a gas turbine to make electricity. The heat from this reaction can then be used to produce steam, which turns a standard steam turbine and creates even more electricity.

Meanwhile, all the CO created in the gasification process is reacted with steam in a water gas shift reactor (is it coming back to you, yet?) to produce H_2 and CO_2. At this point, the H_2 can either be bled off as use for fuel, or sent back to the gas turbine to be burned in the generation of electricity. The CO_2, on the other hand, is destined for a deep, dark grave.

As an added bonus, the gasification process also allows for the removal of sulfur- and nitrogen-containing compounds, which can then be used in the manufacture of fertilizers and other products. For more about this and other coal-related issues, visit the Coal Utilization Research Council website at *www.coal.org*.

Is it really possible to build a squeaky-clean coal gasification plant? Sure, but it's not cheap. In 2003 the DOE announced plans to build a joint public and privately funded 275 megawatt plant over the next ten years, at a cost of $1 billion. The plant is designed to be totally pollution-free. For more information on this proposed plant, visit *www.gcrio.org/OnLnDoc/pdf/power_plant.pdf*.

non-renewable coal

The one insoluble problem with coal, of course, is that it's not renewable. At our present rate of consumption of just under one billion metric tons per year, all the recoverable coal in the United States will be gone in 250 years. And if production is stepped up for hydrogen production to run our cars? It could be gone in half that time.

Clearly, the day will come when we'll have to wean ourselves away from our coal habit, and most would say the sooner the better. But that doesn't mean the baby has to be thrown out with the bathwater. Coal or no coal, gasification is still a cool technology, and it offers a great deal of promise for the future. We'll simply have to be more discriminating about the things we chose to gasify.

biomass option

That's where biomass comes in. It's attractive for a number of reasons, not the least of which is the fact that there are countless millions of unused acres of farmland in the U.S. that are just waiting for the chance to grow crops that turn water and carbon dioxide into high-energy hydrocarbons;

compounds in which the hydrogen is far more accessible than it is in the low-energy substance we call water.

As Zed pointed out, biomass includes everything from carefully selected crops, such as switchgrass, to agricultural refuse—ranging from corn stalks to nut shells; to forest refuse and garbage. In the end, market and legislative forces will dictate which sources are the most economically attractive. (For a brief DOE report, titled *Techno-Economic Analysis of Hydrogen Production by Gasification of Biomass*, visit: *www.eere.energy.gov/hydrogenandfuelcells/pdfs/iib5_bowen.pdf.*)

As an erstwhile farmer, I can say with confidence that it would be a farmer's dream to earn money from a crop that doesn't need to cultivated, weeded or irrigated, and doesn't have to satisfy the cravings of someone's—or some thing's—palate.

sequestered carbon used in fertilizer

The idea of sequestering carbon as a charcoal-based fertilizer is the brainchild of Danny Day of Eprida, Inc. (*www.eprida.com*). While charcoal has been used for over 2000 years to enrich farmland in the Amazon basin (the soil thus created is called *terra preta*), the notion of producing the charcoal during the gasification of biomass and then fortifying it with nitrogen captured during the process, is certainly fresh and timely. Day's process—now in the pilot stage—currently produces 8 kilograms of "char" pellets for each kilogram of hydrogen. The process, which reuses 94 percent of the heat produced, requires no fossil fuels. As an added benefit, nitrogen- and sulfur-containing byproducts obtained from the gasification of coal can be used to further enrich Eprida's ECOSS™ (Enriched Carbon Organic Slow-released Sequestering fertilizer). For more information, check out the Eprida website or *www.ases.org/hydrogen_forum03/Day.pdf* on the American Solar Energy Society website.

Of course, when we're talking about ag products, gasification for the production of hydrogen, fertilizer and electricity is just one side of the coin. There are also the alcohol fuels, ethanol and methanol, to consider.

But that's another story...

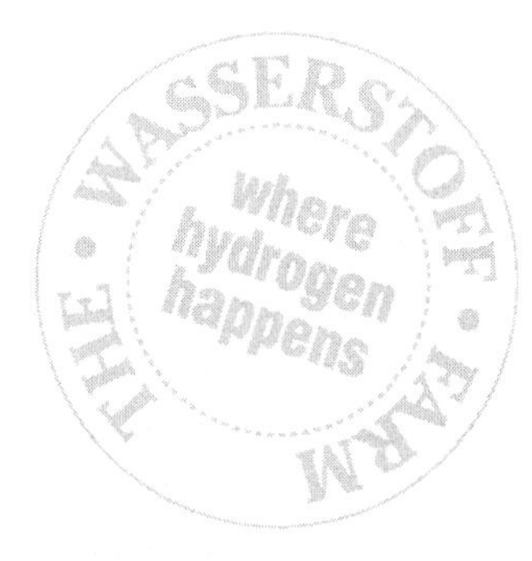

chapter 11

The Alcohol Sisters: A Hot New Fling for Hydrogen?

Zed was nowhere to be found the next morning. I checked the greenhouse and the holosium inside and out, and everywhere else I could think to look. There wasn't a trace of Zed. This was unusual, to say the least—when he wasn't waking me up at the crack of dawn, he was usually just outside waiting impatiently for me to flush the sleep gremlins out of my brain with a flood of hot coffee. Could something have happened to him?

Maybe he was at Wasserstoff World. Slowly I pulled open the door to the small building that housed the dimensional compressor and peered inside. Sensing my presence, the eerie contraption turned itself on, opening a shimmering black doorway to...? A shiver ran up and down my spine; I didn't trust the thing even when Zed was at the controls. Without Zed, who knows where I'd end up? For all I knew, he'd gone to the moon in the middle of night and hadn't reset the thing.

So I just had to keep looking for him. To cover more ground, I saddled my horse, who, as I quickly noticed while tightening the cinch, was as full as a tick from his 24/7 job as Zed's lawnmower. Just the same, he was still young and frisky and seemed happy to have the exercise.

With a horse under saddle and Zed strangely absent, the thought of going deeper into the Wasserstoff Farm was tantalizing. I knew in my bones a secret was lurking there, and I wouldn't be satisfied until I'd uncovered it. And why not? I was just looking for Zed, right? I had to see if he'd been gobbled up by the Wasserstoff Monster.

I made a beeline toward the center and...quickly found Zed, instead.

He was hidden in the middle of a thick stand of juniper, operating some strange contraption with a bunch of copper tubes and stainless steel vessels, one of which was set up over a burner giving off a light-blue flame. Three larger vats sat in a row off to the side, all propped-up on cinder blocks. One of them was heated with a flame from beneath.

"Hey there, Zed!" I said in greeting. As usual, he was dressed in jeans and a long, heavy lab coat. Today's T-shirt read: *The Fuel's Homegrown at the Wasserstoff Farm*. A long stalk of grass hung casually from the side of his mouth. On his head he wore a straw hat that looked as though it might have been in someone's crosshairs once or twice.

Quickly he jumped up and looked around suspiciously in all directions, then fixed his narrowed eyes on me, saying, "Y'all didn't bring no revenuers with ya, didja?" So *that* was it—Zed was running a still.

Still atop Mike, I answered, "C'mon, Zed! You know there's no one here except you and me...and whoever else you have hidden around to take care of the place. How could a 'revenuer' get in here, anyway?"

Still vigilant, he said, "Ya never know. Them boys is *sneaky*."

"What are you worried about?" I wondered, "you're making this stuff for fuel, aren't you?"

"Well, a'course I am, but I ain't got no license, and them Treasury fellas ain't got no sens'a humor."

"Really, Zed. I'm sure no one's here but the two of us," I assured him.

With that, he threw his hat into the bushes and spit the grass stalk out of his mouth. And, fortunately, he also cast off his hayseed demeanor. "Good," he said, relief evident in his voice, "a person can never be too sure of these things, especially when he's dealing with the 'Big G'."

Sliding off Mike's back, I asked, "What's with the hat and the cracker accent, anyway?"

"A disguise, of course. You don't think I'd go to jail as who I really am, do you?"

Actually, I mused, Zedediah Pickett was a great name for a backwoods moonshiner, but I wasn't about to point that out to Zed. Poking fun at a man's name was real close to the top of the *things-never-to-do* list; right there between *don't use gasoline to start a fire in your wood stove*, and *don't tether a fractious horse to the side mirror of the local sheriff's car*. So instead I said, reassuringly, "Really, Zed, I'm sure I wasn't followed. Why don't you quit worrying and show me how this thing works?"

And so he did.

"Well, okay. You see, I'm making ethanol, otherwise known as grain alcohol, or ethyl alcohol. Or simply '200 proof shine'. The process is a wonder of biological production. Chemically, ethanol is C_2H_6O, so you can see it's got a good 3:1 hydrogen to carbon ratio. But it doesn't start out that way, of course."

He walked over to a large vat, heated by a slow light-blue flame which, I guessed, was burning hydrogen gas. Zed slid back the lid to expose the contents. It looked like nondescript gunk, and had a peculiar odor that was at once sweet and sour. Zed replaced the lid, and said, "Here I'm cooking the mash with a special enzyme to liquefy the starch. I make my mash from ground-up corn, like most big-time ethanol plants, but these days it can be made from almost anything."

"You mean biomass?"

"Exactly. It all works, but corn is the easiest."

The next vat was open; its contents were milky, in an ugly sort of way. "After the starch is cooked into a gummy liquid, it has to be converted into sugar. In the old days, the moonshiners would first sprout the corn by keeping it warm and moist for several days. It's the natural way to make sugar from starch, but it's much faster and efficient to just add enzymes to the mix."

"What's the difference between sugar and starch, anyway?" I asked.

"Not much, really. Sugar is to starch what a single link is to an entire chain. Starch is an efficient way for plants to store energy, since it can easily be pulled out of storage and burned as fuel, simply by breaking a few bonds.

"Of course, when you start talking about using biomass, you are, by and large, talking about breaking apart cellulose which, interestingly enough, is similar to starch in that it is a chain of glucose molecules. The difference is that the glucose—or sugar—molecules attach to one another with hydrogen bonds—as in that intractable molecule, water—and so it's a bit more of a chore to break cellulose back down into sugar. Fortunately, it can be done with enzymes too."

Pointing to an open vat with a bubbly layer of brown scum floating on top I asked, "What's in the third vat?"

"This is the fermentation vat. It's where yeast turns the sugar into alcohol and CO_2."

"How does *that* work?" I wondered.

Quickly he said, "Trust me, you don't want to know. It's a terribly complex twelve-step process called 'anaerobic respiration', which simply means energy is being produced in the absence of oxygen. It's the exact same process in yeast and humans, right down to the end products, which are ethanol and CO_2 for the yeast, and lactic acid for us."

"Lactic acid?"

"Right. Here's how it breaks down..."

I expected Zed to make those rod-like letters appear in the air, as before, but this time he simply waved his arm in an arc, making the natural background disappear into a coal-black void. Then, with an artful finger, he wrote:

$$C_6H_{12}O_6 \textbf{ (glucose) + botanical miracle -> 2 } CO_2 \textbf{ + 2 } C_2H_6O \textbf{ (ethanol)}$$

Wherever his finger traced a line through the curiously black background, trees and sky and light appeared, as though he were scratching soot from a blackened window.

"That's what happens with yeast. With people the reaction is like this:"

$C_6H_{12}O_6$ (glucose) + grueling exercise -> 2 $C_3H_6O_3$ (lactic acid)

"Lactic acid looks like half of a sugar molecule," I observed.

"So it does. But it's not much good for anything besides making your muscles burn. The body does, however, turn lactic acid back into pyruvic acid, where it's used aerobically when oxygen is available, so it's not really waste.

"On the other hand, the ethanol and carbon dioxide made by the yeast are indeed wastes—they're what's left after the energy has been taken out of the sugar, and they are of no further use to yeast."

He waved his hands and the virtual chalkboard vanished. "But that's neither here nor there. By the time our fermentation is finished, all the sugar is turned into ethanol and CO_2, and all we have to do is separate them from the water, the mash, and the spent yeast. The CO_2, of course, bubbles out during fermentation—where, by the way, it can easily be captured—while the ethanol needs to be boiled out of the mix and condensed."

"That's the condenser?" I asked, pointing to a coil of copper pipe running through a large vat of cold water.

"Yep. By the time the ethanol works its way through the condenser, it's close to 200 proof. In a commercial operation it would then go through a drying process where the remaining water is removed. Of course, if all you want is shine, 200 proof doesn't leave a lot of room for flavoring. It's best to make it weaker and get a little creative with the ingredients."

With that, he took a dirty old cup and dribbled a few drops into it from a petcock near the bottom of the condenser. He sniffed it and threw it back hard, then screamed, "Whoooooeeee! That's *fine* shine!"

Zed sez

Ethanol solves the problem: it's already a liquid. It's easy to handle and it's not nearly as toxic or hazardous as gasoline. And, if you run it through a fuel cell instead of an internal combustion engine, it's extremely efficient.

Calming my horse, who had taken immediate exception to Zed's strange behavior, I said, "Zed, is there something in your past I should know about?"

He stiffened, and answered indignantly, "What *ever* do you mean?"

"Oh, never mind...I just thought, well..."

"Thought *what?*"

His hands rested on his hips as he thrust his chin forward. I could see I was getting myself into trouble. Quickly I said, "I thought that, uh...it looks like a lot of work making shine, and it's amazing you should know so much about it."

He retorted, "That's my job. To know stuff. So I can teach it to ninnies like you."

Letting his last remark glance off as best I could, I tried to make the subject veer to one side or the other. I waved an arm at his still, and said, "Well, then why don't you explain how this contraption for making fuel could possibly be productive."

"This? Get serious! This isn't good for anything but making a few quarts of joy-juice every so often. It's nothing like the slick corporate operations in the grain-growing states. Some of those facilities can turn out 100 million gallons of the stuff per year. You pour ground corn in one end, pull ethanol out the other. And to sweeten the deal, you get corn oil and high-protein food and feed additives as byproducts."

I said, "That's really good, Zed, but aren't we here for the purpose of getting this hydrogen thing kick-started? What do we accomplish, really, by turning corn and grass and ground-up trees into a kinder and gentler form of automotive fuel?"

He dispensed another swallow of shine into his cup, and said, "I was just getting to that. You see, you can think of ethanol as just a cleaner, though less energetic, replacement for gasoline, in which case the atmosphere becomes less of a greenhouse and we go about doing things as we always have."

He rolled the shine around in the bottom of his cup, then took a sip. He managed to swallow it without whooping or grimacing, and added, "Or you can think of it as a hydrogen carrier. Then things gets interesting. Suddenly, we don't have to worry about compressing and storing hydrogen gas, or figuring out new, efficient ways to liquefy it. Ethanol solves the problem: it's already a liquid. It's easy to handle and it's not nearly as toxic or hazardous as gasoline. And, if you run it through a fuel cell instead of an internal combustion engine, it's extremely efficient."

"But it has *carbon* in it," I objected.

Throwing up his hands and almost spilling his shine, he raved, "Who cares? If you recall, it has less carbon than the sugar you started with. And it's all atmospheric carbon, not subterranean. So, if you sequester the carbon dioxide given off during fermentation, you've actually taken more carbon out of the atmosphere than you will put back into it by driving your car down the road."

It didn't seem that Zed was telling the whole story. "What about all the carbon from fossil fuels you use to produce the ethanol...farm machinery, distilling equipment and so on?" I asked.

"Well," he answered, "we're back to the same problem we have making hydrogen. Do you wait until renewables have come of age, or do you do the best with what you have to get the ball rolling? Most of the power to make shine...uh, ethanol, comes from coal and natural gas. Sound familiar? But, believe it or not, these technologies are getting cleaner."

He drained his cup without making a face, and added, "As we speak they're working on ways to fuel the process of making biomass ethanol by combusting the lignin that can't be converted to starch. Burning lignin is a CO_2 neutral process. It's all going to come around, sooner or later, lad. There are too many clever people working on the problems for it not to."

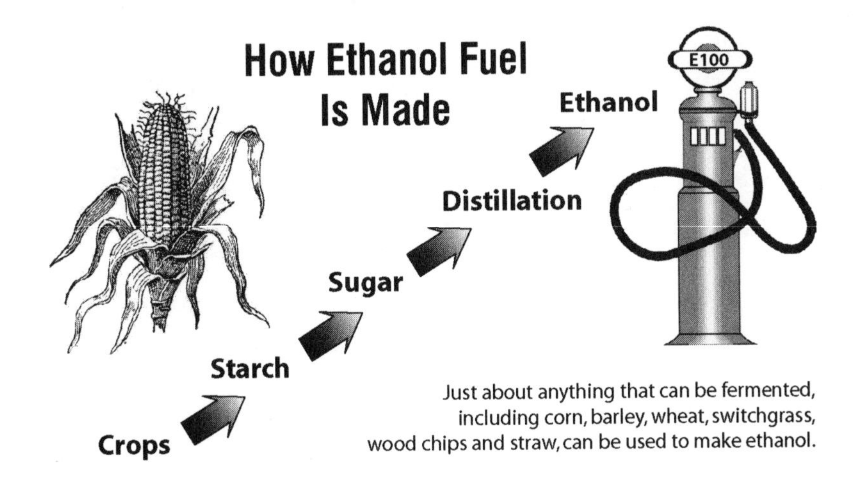

Just about anything that can be fermented, including corn, barley, wheat, switchgrass, wood chips and straw, can be used to make ethanol.

Hoping to pin him down, I asked, "So you think ethanol is the answer?"

Zed bent down and drew another gurgle of shine into his cup. "Depends," he said, "on what the question is. If you are referring to the *big* question of what will be the energy carrier of the future, you need to understand that there is no *single* answer. There will be a place for hydrogen gas, liquid hydrogen, ethanol...."

"And methanol?" I wondered aloud.

"Been doing your homework, I see."

"Doesn't mean I'm real clear on the difference between the two," I answered.

"That's what I'm here for boy, to make you clear."

"Okay. Do it," I said, gazing into his bloodshot eyes.

He took the slightest bit of a sip of shine, smiled brightly, and said, "First off, even though it's alcohol, methanol is a deadly poison. Second, you can't make it the same way you make...uh, ethanol."

"So how *is* it made?"

"Well, basically, methanol is methane with one lonely oxygen stuck between the carbon and one of the hydrogen atoms. That makes it a liquid. It also makes it convenient to handle. As far as producing the stuff, it's made from syngas, which—as I'm sure you recall—is H_2 and CO. So it can be made from coal, natural gas, biomass...anything you can make syngas from."

"But how does it differ from ethanol?"

"Structurally, it has one less carbon and two less hydrogens. That gives it a 4:1 hydrogen to carbon ratio, instead of the 3:1 ratio of ethanol, as I said. That's good; the less carbon the better, from a fuel-cell's point of view."

I scratched my head, and said, "I'd ask you for the bottom line, but I'm sure you'll say there isn't one."

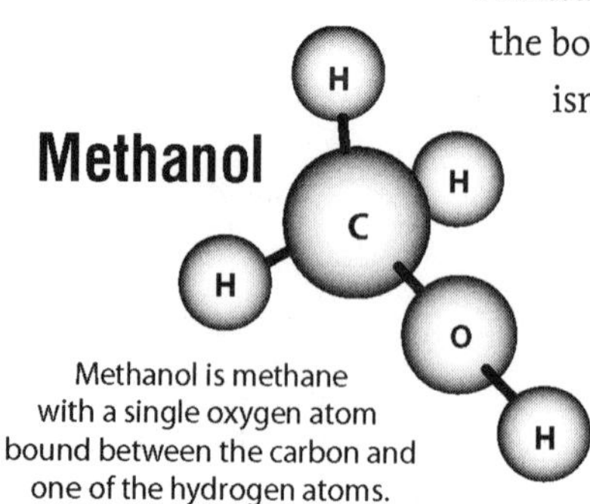

Methanol is methane with a single oxygen atom bound between the carbon and one of the hydrogen atoms.

"Of course there is, lad. But it moves day to day. It all depends on what happens in the future—which fuel automakers choose for their engines and fuel cells, and what new and efficient methods are developed for manufacturing each type of alcohol. And even which industries governments choose to subsidize. But that's politics and I'd sooner teach cats to do tricks than get into *that*, thank you."

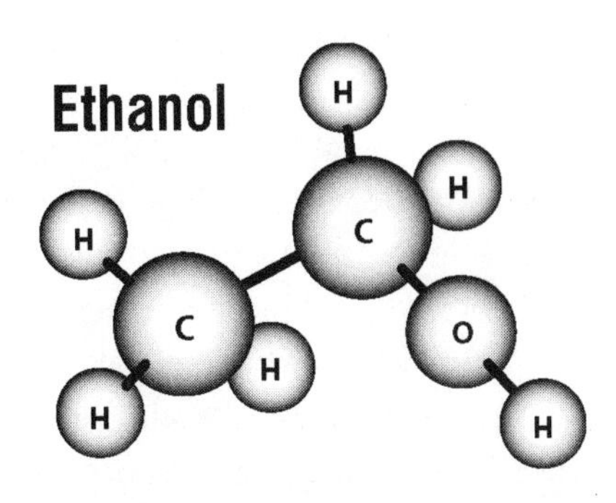

Ethanol is structurally similar to methanol except for a methyl (CH_3) group in place of one of the hydrogen atoms.

With that, he drained his cup, fished his hat out of the bushes, and said, "Are we done? I think I need a nap."

His eyelids were beginning to droop, and his eyes had taken on a noticeably glazed aspect. He sure looked like he needed a nap. "Sure, Zed. But before you do, tell me one thing—where did you learn how to make shine, really?"

He quickly perked up and roared with laughter, scaring my horse again. "The truth? I learnt it from my pappy, who learnt it from his pappy...right on down the line. We Picketts is'a been makin' shine for as long as they's been corn to make it with."

Having said his piece, he lay back with his head against the fermentation vat and nodded off to sleep with his straw hat over his eyes.

Zed sez

Methanol is made from syngas, which is H_2 and CO. So it can be made from coal, natural gas, biomass... anything you can make syngas from.

In that moment I suddenly realized I had: one—a lot of free time; two—a saddled steed; and three—no one around in any condition to keep an eye on me. And, of course, I had an unquenchable curiosity about what lay in the center of the Wasserstoff Farm.

Although I was feeling neither as determined as Odysseus, as righteous as Lancelot, nor as deluded as Don Quixote, I set out on my quest. It was a quick ride through the ponderosa forest and into the orchard, and as before, the light breeze that originated just inside the wall grew steadily stronger as we moved closer toward the center. And, as before, it grew progressively hotter.

At the edge of the orchard Mike and I stopped to share an apple that I halved with my pocketknife. In truth, I'd about had my fill of fresh fruit, but a plump apple dangling from low-hanging branch was a temptation I could not easily resist. Mike quickly downed his half and began eating grass, while I nibbled on my half and studied the situation.

As before, the high-pitched whine emanating from the interior made Mike a little edgy, which was natural. A horse that likes wind has yet to be made, but, to Mike's credit, he seemed to be getting used it. Between the two—the strange, distant machinations and the steady wind, I would say the wind was the most disconcerting. It seemed to come from all directions, or more correctly, always from the perimeter toward the center.

But why? It made no sense.

I touched Mike's flanks with my heels and headed him into an endless wheat field, its shimmering heads of grain slowly undulating like waves on a velvety golden ocean. With each step the ubiquitous whine grew in intensity, and the wind stiffened. After perhaps a mile the field ended as the stalks of wheat grew shorter and shorter, until they gave way to a windswept no-man's land, where only short grasses grew.

The heat was becoming oppressive; I wiped the sweat from my brow with my sleeve.

In the distance—perhaps another mile or more—I could see, ghost-like through the dust and haze, a large, dark structure rising from the ground. It must have been the center of the Wasserstoff Farm and the source of the noise, which now gave off a low, drumming sound, underlying the relentless whine.

I coaxed Mike forward, but after just a few steps he planted his feet and refused to move. All my attempts to get him to go further were answered with increasingly belligerent refusals. "C'mon, Mike," I pleaded, stroking his neck, "let's get to the bottom of this."

But Mike just snorted and jittered, and gestured that he was in the mood for a rodeo, if it came to that. And I was of no mind for hoofing it on my own. If I wanted to see what lay at the heart of the Wasserstoff Farm, I would have to find another way.

I turned the horse around and headed back into the wind.

technistoff - 11

There's a subtle but unmistakable irony in the thought that a substance with such a colorful and sullied past—thanks in no small part to Zed's ancestors—might end up being the energy carrier of the future. Ethanol isn't shine, of course, but it's not too far removed from it. The science of mass-producing ethanol and the art of making shine share as many similarities as they do differences. The main difference, of course, is in the end product. Since the moonshiner wants a product that retains the flavor of the mash, the ingredients are carefully selected and the distillation process terminated when the shine reaches about 100 proof, or 50 percent alcohol. The ethanol producer, by contrast, couldn't care less about flavor. The name of the game is to churn out pure alcohol in volume, so anything that can be turned into sugar will work. The cheaper the better.

Currently, there are more than 90 ethanol production facilities in the USA (see page 141), with a total capacity of over 3.6 billion gallons per year (*www.ethanolrfa.org*). How do they do it?

6-step process of making ethanol

Producing ethanol from corn, the primary feedstock used by large-scale ethanol producers, is a six-step process. (Corn has traditionally been the crop of choice because it's fast growing and loaded with starch.)

1. **Milling**—There are two types of milling processes: dry and wet. In dry milling, the corn is simply ground into meal by running it through a hammer mill. In the wet milling process, the corn is first steeped in a hot sulfuric acid solution for one to two days, to extract the corn oil before it is dried and ground, as in the dry milling process.

2. **Liquefaction**—During liquefaction the corn meal is mixed with water to which an enzyme, alpha-amylase, has been added to turn the starch into a liquid. It's done in large pressure cookers at 120–150 degrees Celsius. As Zed pointed out, the old-timers simply sprouted the corn to coax it into making its own sugar. A rather slow and inefficient process, but clever.

3. **Saccharification**—This is the part where saccharin is added. Just kidding. It's really the stage where the mash is cooled and another enzyme, gluco-amylase, is added to convert the starch into dextrose (glucose).

4. **Fermentation**—Fermentation is the magical step where the yeast eat the sugar and excrete ethanol and CO_2. The CO_2 bubbles to the top and is easily captured. In modern facilities the mash is moved continually through a series of fermenters, so new mash is always coming in one end, while alcohol-laden mash (about 20 proof) spills out the other. The yeast don't seem to mind being jostled about.

5. **Distillation**—Up to this point all you've made is corn beer, and if you did it on a small enough scale, Uncle Sam wouldn't mind. But once you start distillation, you'll need either a license, a good lawyer, or a dark place to hide. In modern distillation plants, the mash is pumped through a multi-column distillation unit until it reaches about 96 percent purity (192 proof).

6. **Dehydration**—Dehydration removes the last of the water. It's done by running the vaporized ethanol through a molecular sieve—a bed of ceramic beads that absorb the remaining water. Once past the sieve, the ethanol is 100 percent pure. However, to make sure no one drinks the stuff, it is then "denatured." This is a euphemism for "poisoned," usually with gasoline.

This description of the ethanol production process was loosely adapted from one provided by East Kansas Agri-Energy at: *http://www.ekaellc.com/process/*. But if big production plants are boring and you'd rather read a colorful little treatise on making shine, visit: *http://smokymtns.com/dew.htm*.

efficiency of making ethanol

How efficient is this process? Well, from every bushel of corn, 2.7 gallons *[10.2 liters]* of ethanol can be produced, along with 17.5 pounds *[7.9 kg]* of distillers dried grains and 17 pounds *[7.7 kg]* of carbon dioxide (*http://www.bioproducts-bioenergy.gov/exist site/pdfs/drymill_ethanol_industry.pdf*). From 1997–2002, U.S. farmers averaged a little over 132 bushels of corn per acre (as provided by the Missouri University Food and Agricultural Policy Research Institute: *www.fapri.missouri.edu*). That yields about 356 gallons of ethanol per acre, on the high side.

But how much energy did it take to make that much ethanol? In a 2002 paper titled, *Ethanol Energy Balances*, prepared by David Andress, of David Andress & Associates Inc., for the U.S. Department of Agriculture (see *www.afdc.doe.gov/pdfs/6865.pdf*), Andress concludes that the net energy gain for the production of corn ethanol is between 21 and 34 percent. This means that after accounting for all the energy used in production, you would still have a net gain of between 75 and 121 gallons *[284 - 458 liters]* of ethanol per acre.

energy required to make ethanol

It doesn't sound like much, but when we compare it to a *really* inefficient process, like using electrolysis to get hydrogen from water (postulating an 85 percent electrolysis efficiency, and a 50 percent hydrogen-to-electricity efficiency), it would require using the energy equivalent of 100 gallons *[378.5 liters]* of hydrogen for every 42.5 gallons *[161 liters]* of hydrogen produced. It seems like a paltry amount, I know, but at least there are numerous ways in the works to dramatically improve the dismal numbers for electrolytic hydrogen.

By contrast—according to Andress—gasoline is stuck at a minus 19 percent net energy loss, meaning that it takes the energy contained in 100 gallons *[378.5 liters]* of gasoline to make 81 gallons *[306.5 liters]*.

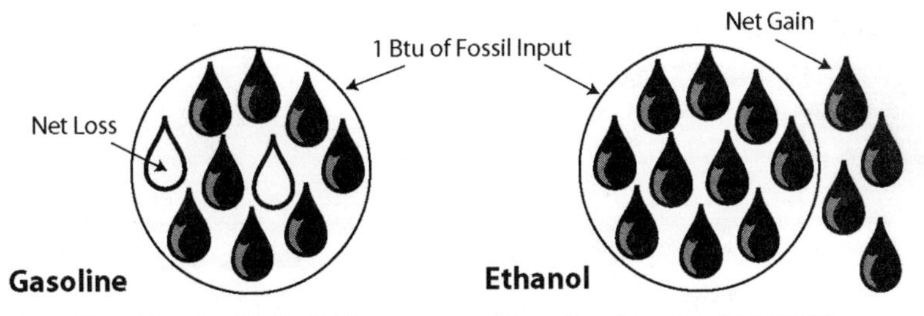

Energy Yield of Gasoline vs. Ethanol

Source: Argonne National Laboratory and www.ethanolrfa.org

feedstocks for ethanol

But back to ethanol, and what we can do to make it even more appealing. To begin with, we need to remember that corn is not the only possible feedstock for producing ethanol. By using a cellulosic feedstock such as switchgrass—a tall, hearty, fast growing perennial grass that once covered much of the Great Plains—instead of corn, the efficiency of the ethanol production process can be increased several-fold.

How? For starters, the ethanol yields per acre of switchgrass can easily be double those for corn: 350 gallons per acre for corn versus 500–900 gallons per acre for switchgrass, according to *Biofuels from Switchgrass: Greener Energy Pastures*, a 1998 paper prepared for DOE's Office of Transportation Technologies, and available at *http://bioenergy.ornl.gov/papers/misc/switgrs.html*. That means, on average, we can get twice as much ethanol per acre by growing a plant that doesn't need to be seeded every year, or pampered in all the ways corn is pampered. A lot of energy can be saved in planting and harvesting switchgrass, but that's really a small part of the equation (about 0.09 Btu of petroleum used for every Btu of ethanol produced, according to Andress). Essentially, higher production just means more ethanol; it says nothing about processing efficiency.

To increase efficiency, a novel way of powering the production plant is needed. And therein lies the beauty of cellulosic ethanol for, while cellulose and hemicellulose are made of starches and can be enzymatically treated to produce sugar, the lignin present in plant tissues is not made of starch and cannot be broken down into anything that yeast would consider food.

So, if the yeast won't eat it, why not burn it? Andress suggests that ethanol production plants can be built that use lignin—recovered from the distillation process—to generate all the electricity and heat needed to produce ethanol.

Switchgrass is an ideal energy crop. It can grow to 10 feet high in a good growing season with stems as strong and thick as hardwood pencils. It uses water efficiently, and grows fast, capturing lots of solar energy and turning it into chemical energy—cellulose—that can be liquified, gasified or burned directly. Expected yields in field-scale production are in the range of 5 – 9 tons per acre, enough to produce 500 to 900 gallons of ethanol per acre per year.
Source: Bioenergy Feedstock Development Program, Oak Ridge National Laboratory, Tennessee.

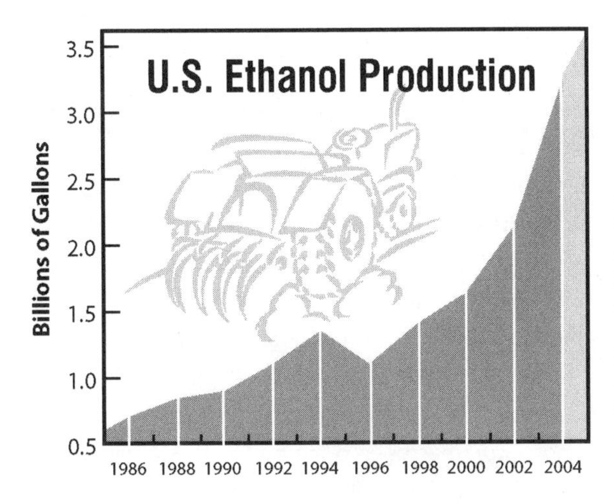

Ethanol production facilities in 20 states will produce over 3.3 billion gallons of ethanol in 2004.

Feed stocks include corn, barley, brewery waste, cheese whey, and potato waste.

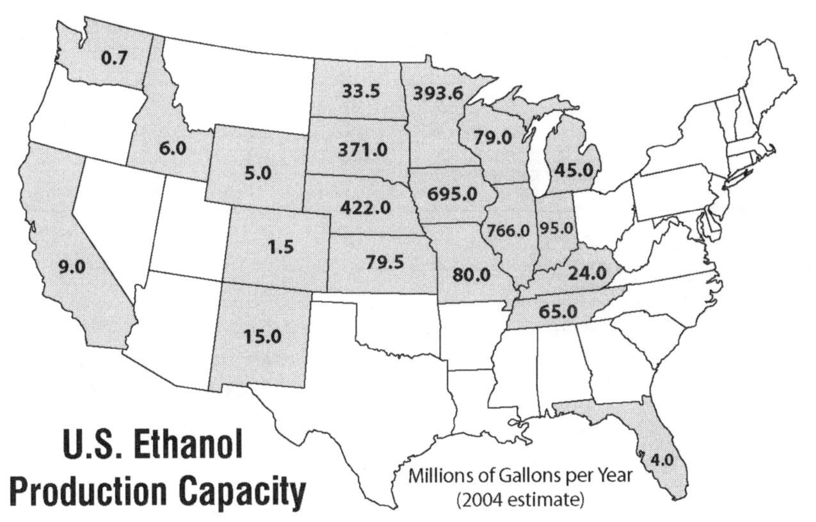

These plants would emit no net greenhouse gases. Suddenly, the net energy gain has skyrocketed to over 80 percent. For every 100 gallons of ethanol produced, we own 80 gallons free and clear, energetically speaking.

methanol facts

Where does methanol fit into this scheme? Traditionally, ethanol and methanol have not had to compete for feedstock. Ethanol has always been made from corn, sorghum and barley, while methanol is customarily produced from syngas (CO and H_2) derived from coal and natural gas. But now that biomass has proven itself a promising feedstock for both, the alcohol sisters find themselves in a hair-pulling contest. Who will prevail? As Zed pointed out, it will largely depend on legislation, technology and market forces, as well as other untidy considerations, such as capital costs.

Still, a couple of points should be made. First, methanol has only 75 percent of the energy density of ethanol: 10,250 Btu per pound *[23,840 kj/kg]*, compared to ethanol's 13,160 *[30,610 kj/kg]*. That means a tank holding enough methanol to take you 75 miles down the road would take you 100 miles if it was filled instead with ethanol. And it doesn't matter if you were driving a car powered by an internal combustion engine, or a fuel-cell stack and electric motor. Energy is energy.

On the other hand, more work has been done on Direct Methanol Fuel Cells (DMFC) than on Direct Ethanol Fuel Cells (DEFC), and if there weren't armies of researchers working feverishly to change that fact, it would appear that methanol might win the day.

But it's going to be a long day, and it ain't even half way to noon yet. So grab yourself a cup of shine...uh, iced tea, I mean. Then sit back and enjoy the race to make the world a cleaner place.

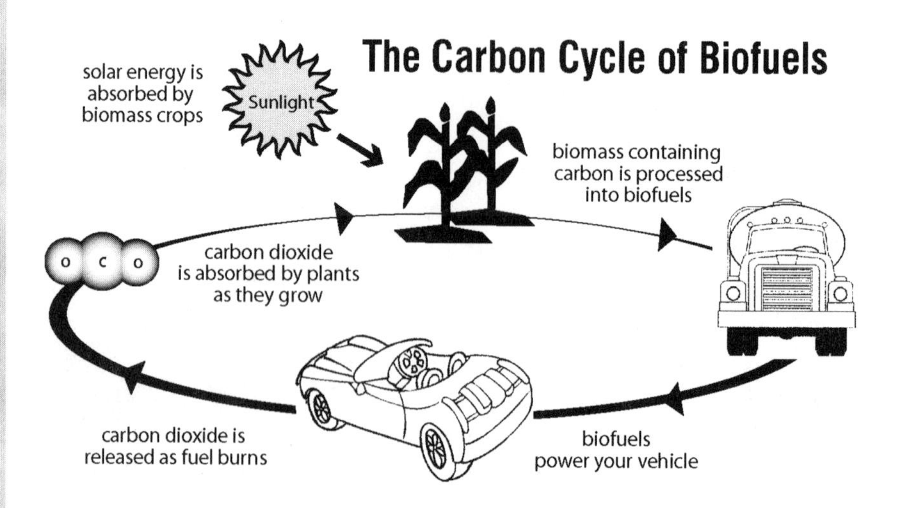

chapter 12

To Nuke or Not to Nuke: Is That a Question?

When I returned from my unsuccessful attempt to reach the heart of the Wasserstoff Farm, I fully expected to find Zed still soundly asleep against the fermentation vat. Instead, he was gone; him and every last trace of his illegal moonshinin' operation. The grass wasn't even matted in his little hideaway among the junipers. It was like he'd never been there.

I rode around the area several times, just to assure myself I was looking in the right place. I was. All the rocks and trees were where they should have been. Zed and his still were just plain gone.

So where was he now?

I decided to ride back to my cottage and wait. If I was lucky, I might be able to catch a little nap before Zed came knocking on my door. That would be nice, I thought; just a short rest before...*what the heck?*

Daydreaming about the possibility of a few minutes of midday shuteye, I must not have been paying any attention to where I was going. And now I was...nowhere. Literally. Mike and I had ridden into a perfectly featureless landscape. A ubiquitous white brightness bathed the sky, the earth beneath Mike's feet, and the strangely-absent landscape for as far as the eye could see in every direction. It was like being inside a giant lightbulb.

Oh, my god, Zed, I thought, *what have you gone and done now?*

Oddly, Mike was not in the least bit perturbed by the lack of...everything. He seemed curiously calm, as if inside a soft and pleasant dream. I myself felt as though me and my inexplicably mellow horse were the only two things remaining in the universe, and it was hardly a comforting thought.

Presently, something new appeared in our vacuous micro-universe. About the size of a basketball, it was a tight snarl of little blue and green balls; less than a hundred of the blue ones, but well over a hundred of the green ones. It came to rest about 50 feet in front of us, waiting. As innocuous as it seemed, I began to feel uneasy—I remembered those balls from a couple days ago. The blue ones were protons, the green ones neutrons. I was looking at the nucleus of a very heavy atom. The whole situation had an ominous feel to it.

Suddenly, from out of the blue—or, I should say, white—a single green neutron crashed into the nucleus like the cue ball slamming into the rack on a break. Several things happened very quickly after that. I felt a great rush of heat and energy as two or three neutrons were liberated and went zooming off into space. The remaining neutrons and protons broke into two new nuclei, and they, too, gave off an uncomfortable surge of heat before settling down into a pair of uneasy nuclei.

My horse was starting to toss his head, and shift his weight from foot to foot. If I got unseated because of one of Zed's overly-dramatic demonstrations... "Zed?" I hollered, turning Mike in all directions, trying to catch a glimpse of the old wizard. "Where are you, and what in the world are you doing?"

"Quit whining and pay attention," came the terse reply—from all directions. Zed was nowhere to be seen, but his voice sounded as though it were coming from inside a movie theater with Dolby surround sound. "What you just witnessed was the fission of a single atom of U-235," Zed's voice boomed through the void. "The energy released was on the order of 200 million electron volts."

"I don't care if it was on the order of the King of Sweden, Zed. You're scaring my horse!"

"And you call yourself a horseman!" The voice was normal, this time, and right behind me. I spun Mike around to see...Zed. He was wearing a new T-shirt; this one said, *Nuke'em, Dan-O*. He reached out, patted Mike on the neck, then whispered something in his ear. Instantly I felt all the tension go out of the colt's body. "There. Your horse is happy, again. Can we get on with it?"

Mike might have been happy, but I wasn't. "Get on with what? I don't even know where we are, Zed!" I replied, sliding off Mike's back.

He scanned me with intensely clear blue eyes. If he had a hangover, it didn't show. "What's this obsession you have with always needing to know where you are? Why not be happy knowing where you aren't?"

"Like where?" I wondered.

"How about the holosium? You seem to have developed a real aversion to that place, so I had to improvise. But enough explaining. Do you want to learn about atomic energy, or not?"

Honestly, I hadn't thought about it. "What's it got to do with a hydrogen economy?" I asked.

He answered, "Maybe nothing. Maybe a lot. Who knows? Just in case, I think it would behoove you to know a thing or two about it."

Sounded reasonable, but I was still leery. Zed had a nasty habit of trivializing the intensity of his demonstrations. "No bombs?"

"Not a chance. The stuff we're dealing with is only 4 percent enriched. That means only 4 atoms out of every 100 is U-235, the fissionable isotope of uranium. The concentration is too feeble to be coaxed into an explosive reaction. The other 96 percent is U-238, an isotope not given to splitting in two when an errant neutron slams into it. Bomb-grade uranium, by contrast, is around 90 percent U-235. Understand the difference?"

"Sure," I answered, uneasily. At least I thought I did.

"Good. Now what did you just see?"

"Well, a green ball—a neutron, I imagine—collided with the U-235 nucleus and broke it in two. A couple of neutrons broke out of the pack and flew off into space. Then it got real hot around here."

Zed nodded. "The heat you felt was a small amount of matter being converted into energy. It's what drives a nuclear reactor."

I thought about that for a moment, then asked, "but what keeps the reaction from getting out of hand?"

Zed said, "Watch. I'll show you." With a wave of his hand several hundred U-235 nuclei appeared and arranged themselves on a plain about two feet thick within a circle. Collectively, they resembled a loosely-packed, 20-foot-wide Alka-Seltzer tablet. Above them, several dozen evenly-spaced vertical rods—each longer than the pack of nuclei was thick—hovered in the air.

"We'll do this in slow motion, so you can see just what's happening."

"Slow motion is good," I told him.

Zed snapped his fingers and instantly one of the hundreds of U-235 nuclei split in two, sending out three neutrons. Two of these neutrons hit other nuclei, cleaving them. These, in turn, sent out more neutrons. Although it was happening slowly, the heat quickly became intense and it appeared as if very soon all the U-235 nuclei would be cleaved by the preponderance of free neutrons. But Zed intervened. By pointing downward with the index finger from his outstretched hand, the long rods began to descend into the swarm of nuclei. "These are called the control rods," he said. "Watch what happens here." To my surprise, every neutron that encountered a control rod was absorbed, and soon the chain reaction began to fizzle. Lifting his finger slightly, the rods began to rise, and things quickly heated up again. By absorbing nuclei-splitting neutrons, the control rods had the effect of determining how much heat the reaction gave off. Simply by raising or lowering the control rods to different levels, Zed could make the reaction proceed at any rate he chose.

"Wow! That's pretty cool, Zed! Is that really how a nuclear reactor works?"

"That's the general idea. In reality, the uranium is in the form of uranium dioxide, and it's pressed into pellets smaller than the first joint of your little finger. Each of these small pellets will, in its lifetime, give off heat nearly equal to the heat energy of a ton of coal. The pellets are stacked into fuel rods, and the fuel rods are collected into a fuel assembly.

"The reactor core, which houses both the fuel assembly and control rods, is made primarily of graphite, and it's all immersed in water which carries heat away from the reactor, where it's used to produce steam, which turns a turbine generator...so on and so forth. I'm sure you know the story by now."

"Why graphite?" I asked.

Zed answered, "Don't get confused when I tell you, but it's needed to moderate, or slow down, the high-speed neutrons given off during fission."

"I thought the control rods did that," I said.

Zed shook his head. "The control rods, usually made from cadmium or boron, stop the neutrons, cold. The graphite just slows them down."

"So the reaction doesn't get out of hand?" I asked.

Again he shook his head. "No, so the reaction doesn't stop altogether."

"Say what?" My face must've been a study in puzzlement.

Zed laughed, then said, "As counterintuitive as it sounds, high-speed neutrons don't split U-235. Slow-speed neutrons do. That's why they need a moderator—it's the only way the reaction can proceed."

I said, "Okay. So you've got water around the reactor ferrying heat to a steam generator. You've got control rods that absorb neutrons and are capable of stopping the reaction. Sounds simple enough. What can go wrong?"

"A lot," Zed said. "At Three Mile Island a control valve got stuck in the secondary cooling system, and stopped the heat transfer from the reactor. The reactor began to heat up, eventually leading to another valve failure, this one within the reactor core's pressure relief system. The water covering the reactor core boiled off. At that point the Safety Injection System designed to flood the core with water should have turned on automatically, but it was stopped by the operators who, in their confusion, thought the core was still immersed in water. By the time they realized their mistake and flooded the reactor, half of it had melted."

"Wow! It's amazing no one was killed."

Zed said, "Fortunately, the Three Mile Island reactor, unlike the one at Chernobyl, was enclosed within a containment structure designed to block

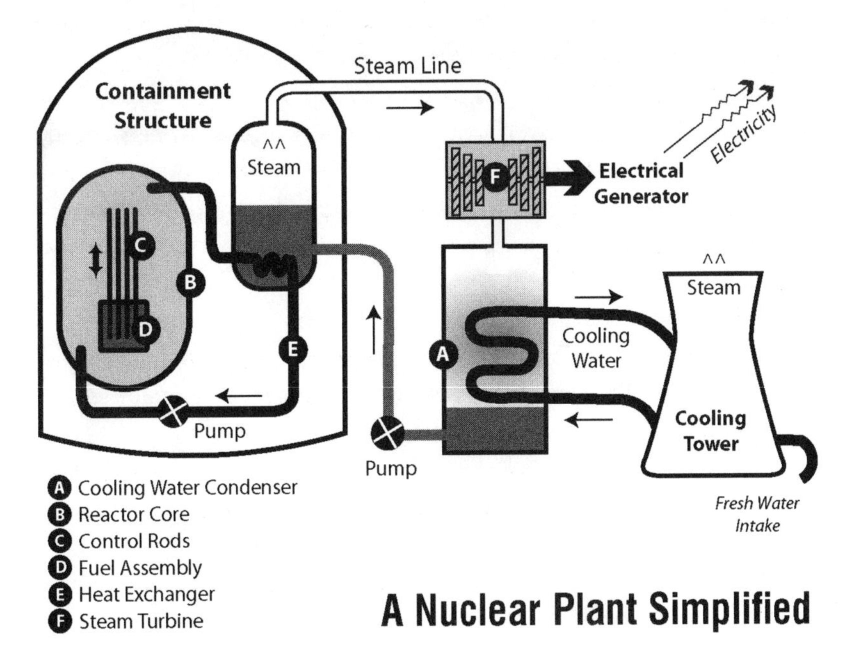

A Nuclear Plant Simplified

the spread of radiation. If there had been a similar structure at Chernobyl, no radiation would have escaped, and perhaps no one would have died."

"So what are you saying, Zed? That you think nuclear reactors are safe?"

He answered, "Safe? Safer than what? Coal mining? Definitely. Solar and wind power? Of course not. It's all a trade-off. The question is, how long do you want wait to develop the technologies needed to get the 'hydrogen economy' into full swing? Twenty years? Thirty? At what point do we trip a trigger in the systems that control the Earth's climate, and suddenly all of the complicated climate-modeling algorithms we've relied on to determine the future of the planet collapse into one colossal '*Oops!*'?"

As much as I hated to admit it, he had a point. I opened my mouth to speak, but Zed cut me off, saying, "On the other hand, you can't just go out and buy nuclear power plant kits and erect them all over the place like so many houses. It takes hundreds of millions of dollars and several years' time to get each one approved, built and online. And by that time who knows if they'll even be needed? The point is, nuclear power is an option that needs to be looked at, like everything else."

"I agree, Zed," I told him, "but maybe we need to look at the advantages and disadvantages that nuclear has compared to coal, its closest competitor at the moment. And for that matter, what each can do to help move the concept of the hydrogen economy closer to reality."

I looked around and noticed that the landscape hadn't changed. "And by the way," I suggested, "might we have a little change of scenery around here? White is nice, but really..."

Zed said, "I'll think about it. I'm not quite done yet. Anyway, your horse likes it."

I looked over at Mike. His eyes were glassy and he was swaying slightly from side to side. I could almost swear he had a smile on his face. "What'd you say to him, anyway?" I asked suspiciously.

"I just offered the poor, misunderstood animal a little horse sense," Zed answered. "Something I'm sure you wouldn't understand."

"Takes one to know one, I guess."

"Humph," he growled.

Better to quit while I was ahead. "Anyway, as you were saying...?"

"Uh, yes. Coal versus nuclear. First off, what part will either or both play in the hydrogen economy? Let's begin with obtaining hydrogen from water. As it stands now, there are two possibilities for giant power plants: thermochemical water splitting or electrolysis. The first option is somewhat similar to the direct solar thermal water splitting we discussed earlier, except in this case the heat would be provided by either fission of radioisotopes or the combustion of coal. In either case, recoverable chemicals would be used to hasten the process. The second possibility, electrolysis, is a process I'm sure will forever linger, fondly and vividly, in your memories..."

"You can say that again, at least the vividly part."

Ignoring me, he continued, "This is assuming, of course, that someone actually builds an emission-free, coal-fired power plant, and it works as efficiently as it's designed to."

"Is this a big assumption?" I asked

"Probably not. It'll get off the ground, sooner or later. Will it be too little too late? Or is this even the direction we want to go? No one really knows, this early in the game."

He raised a finger in the air, and said, "Of course, there is one advantage coal has over nuclear, at least in an advanced coal gasification power plant. You can divert a portion of the syngas and use it to produce pure hydrogen. It's a more efficient process than electrolysis, and you can't do it with nuclear—obviously. The downside to this scheme is that you're getting your hydrogen from a non-renewable source."

"What are the advantages nuclear might have over coal?" I asked. Zed didn't appear to be listening. He was rubbing Mike's nose and whispering some silent incantation. I looked closely at my horse. No doubt about it—Mike was smiling. Zed had put the simple-minded beast under some sort of spell.

"Why don't you tell me?" Zed said, never taking his eyes off Mike.

"Okay, let's see...well, if I remember, you get 2 pounds of carbon

Zed sez

Of course, there is one advantage coal has over nuclear, at least in an advanced coal gasification power plant. You can divert a portion of the syngas and use it to produce pure hydrogen.

dioxide for every pound of coal. So for every ton of coal you transport in, you are transporting 2 tons of carbon dioxide out. Whereas for nuclear..."

Zed appeared to be done placing a hex on my horse. He picked up my thought, saying, "...Your waste is no heavier than your fuel. And far less voluminous. Rather than carting off a truckload of compressed carbon dioxide gas, you're dealing with a spent fuel pellet less than half the size of a cocktail weenie."

"A most unsavory cocktail weenie."

"True. In the end, it depends on how desperately you want the power. Or the hydrogen it can produce."

I said, "A lot of people don't like the idea of getting their energy from such unnatural sources."

Zed raised an eyebrow. "Unnatural? That's a matter of definition. Uranium is more common in the Earth's crust than the tungsten inside your lightbulbs, and the molybdenum in the wrenches you use on your pickup. In fact, uranium is 40 times more abundant than silver. And sooner or later, every last bit of it is going to break down into lighter elements, giving off energy as it does. All we're doing is gathering it together and helping it along. What's unnatural about that?"

"Well..."

Zed patted me on the shoulder, and said, "Oh, I know, lad. It's a dangerous business that's never lived up to its promises. But few things do—live up to their promises, that is."

Somewhat consoled, I asked, "Is there any way to make nuclear energy—and nuclear waste—safer?"

"Sure. It's not like the nuclear boys slinked off and buried their heads in the sand after Three Mile Island and Chernobyl. They went back to the drawing board."

That caught my attention. "And what did they come up with?" I asked.

"A lot. Maybe the most promising idea is the gas-cooled Pebble-Bed Modular Reactor."

"I'm listening."

Zed stood beside me and said, "Alright. Look ahead..."

I did. I saw the familiar control rods, but the fuel assembly had been replaced with solid black balls, like cannonballs the size of tennis balls. There were thousands of them. "Three out of every four of those balls is filled with 15,000 pieces of enriched uranium, each coated with silicon carbide. The remaining ones are made of solid graphite. They..."

"Slow down the neutrons? So the U-235 fissions?"

"Good, lad, good," Zed answered, intent on his work. "The idea here is that the balls—they're called 'pebbles', by the way—can be cycled through the reactor. Fresh pebbles in one end, depleted pebbles out the other. No need to shut down the reactor to refuel. At any one time there are 440,000 pebbles in the reactor, of which 330,000 are fuel pebbles."

As I watched, a multitude of radioactive pebbles slowly rolled—like a herd of round, legless cows—from one end, where they were constantly replenished, to the other, where they disappeared. "What's so safe about this?" I asked.

"A lot of things. For starters, the reactor is cooled with helium, instead of water. Helium, as you no doubt recall, is a thoroughly snobbish gas; it refuses to react with anything else in nature. That means it cannot become radioactive. And it's also non-corrosive. Perfect stuff to cool the reactor and run the turbine."

"Go on..." I insisted.

Zed did. "The reactor is also designed to be 'passively safe'. That means even if the control rods and the helium coolant fail, the reaction slows as

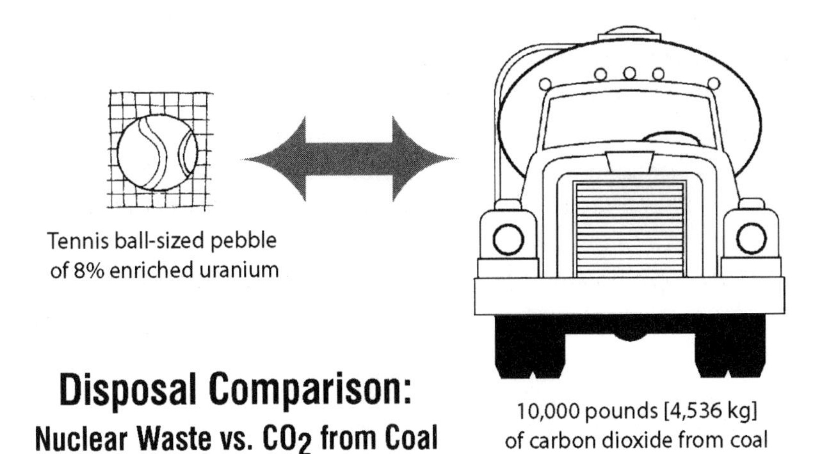

If equal amounts of energy were created in a Pebble-Bed Modular Reactor and a coal-powered plant, you would need dispose of a tennis ball-sized pebble of radioactive uranium versus over 10,000 pounds [4,536 kg] of carbon dioxide from the coal.

the temperature rises. By the laws of physics, the reaction has to stop before the reactor reaches the melting point."

"What about waste?" I asked, chancing to gaze into the vacuous eyes of my bewitched horse, and wondering what Zed had done to him.

"That's an interesting point," Zed said, keenly watching me, as I studied Mike's strange condition. "You see, the pebbles are enriched to 8 percent U-235, and coated with silicon carbide. With that coating they actually take up more volume per unit of energy than 4 percent enriched pellets, but they are designed to keep the radioactive products of the reaction isolated for hundreds of millennia. All that's needed is a place to store them. It's like the problem of what to do with sequestered carbon dioxide, only more, shall I say, critical."

"Sounds like an improvement," I said, warily. "I'm sure there's a downside?"

Zed proclaimed, "Of course there's a downside. The concept is so new it's hard to say if it will live up to expectations. The design is certainly an improvement in terms of safety and efficiency over what's out there now, but, as they say, the proof is in the pudding."

"Might there be other ways to make electricity, and hydrogen?" I asked, in frustration, "where nothing has to be dug deep out of the ground, where carbon dioxide is not produced in copious quantities, and radioactive elements are not concentrated into critical masses?"

Zed snapped his fingers, and suddenly the rocks and trees were again visible. It was good to see colors and shadows, again. The air moved about us, flooding my nostrils with the scent of pine and sage and juniper. Mike shook his head as if waking up from a dream, looked at me quizzically, and then noticed the tall grass tickling his fetlocks. He immediately went to work making it shorter.

Zed strode away, at his customarily brisk pace. Over his shoulder, he said, "But of course, lad. Why didn't you ask sooner?"

I ran to catch up with him, hollering, "Hey, Zed!"

"Yes?" he said, stopping and turning.

"What did you do to my horse?"

"Oh, you'll see, soon enough." And with that he was off.

technistoff - 12

Any day Zed can talk about nuclear energy without singeing my hair, knocking me off my feet, or causing my 1,100 pound horse to have a wildly animated anxiety attack, is a good day indeed. Who says knowledge has to be painful?

isotopes of uranium

There are 16 known isotopes of uranium, ranging in mass number from 242 to 226. But among them—indeed, among all the natural elements—U-235 stands alone. Rather like Zed. As he so vividly illustrated, nuclear reactors owe their success to the peculiar properties of uranium-235, as does the entire science of nuclear energy. All elements heavier than lead have a tendency to decay—usually by emitting alpha particles: two protons and two neutrons; a helium nucleus, in other words—but U-235, with 92 protons and 143 neutrons, is the only naturally occurring element that fissions (splits in two) when struck by slow-moving neutrons. Plutonium-239, the stuff of bombs, also fissions when struck by slow-moving neutrons—much more readily even than uranium-235—but it occurs in nature in such minute quantities that it is, for all intents and purposes, a manmade element, produced by neutron bombardment from U-235's mild-mannered big brother, U-238. Two other fissionable elements exist in nature: uranium-233 and thorium-232, but these two elements are much harder to split than U-235, and it takes fast-moving neutrons to do it.

energy released in fission

As is my habit, I had to verify Zed's claim that the fission of a single atom of U-235 releases 200 million electron volts of energy. To do this, I again consulted Richard Rhodes' monumental work, *The Making of the Atomic Bomb*, and found a number of references that proved Zed true to his word. By contrast, only 5.7 electron volts of energy are released when two molecules of hydrogen (H_2) and one molecule of oxygen (O_2) combine to form two molecules of water (H_2O) in a standard chemical reaction; 1/35,000,000th the energy of a newly-cleaved U-235 nucleus. See the website: *http://hyperphysics.phy-astr.gsu.edu/ hbase/molecule/boneng.html*.

In 1939, German-born physicist Otto Frisch, who later emigrated to the U.S. to work on the Manhattan Project, calculated that 200 million electron volts would be sufficient to make a small, though visible, grain of sand noticeably jump (not far, mind you). That may not sound like much, but if one gram of uranium contains 2.5×10^{21} atoms (2.5 sextillion) and 4 percent of them are enriched (a mere 100 quintillion), then that's a lot of jumping grains of sand. Powerful stuff, that U-235.

I should point out here, for everyone's peace of mind, that 200 million electron volts is only about one thirty-three trillionth of a Btu, so it would take the heat given off in the fission of 33,000,000,000,000 U-235 atoms to equal the heat of one match.

Still, that's nothing to thumb your nose at: 2.5 grams of 4 percent enriched uranium—the weight of a copper penny—contains the Btu value of about 58 gallons *[220 liters]* of gasoline. If your penny was *pure* U-235 (hold the U-238, thank you), it would contain the energy of 1,450 gallons *[5,490 liters]* of gasoline.

Now, to the other side of the coin.

The information available on nuclear energy, I can safely say, ranges from hard scientific fact to pure drivel. I have done my best to steer clear of the latter. But there's no denying it is a hotly debated issue. Some think nuclear power is the neatest thing since the chimney replaced the roof's smoke-hole, while others are convinced it's the devil's handiwork.

Primal fears and knee-jerk reactions aside, there are two ways to view the possible role nuclear energy might play in our future. The most unlikely role—and by far the least imaginative—is just to accept nuclear and use it as an ever-greater factor in our energy future. Build more and more plants until coal is

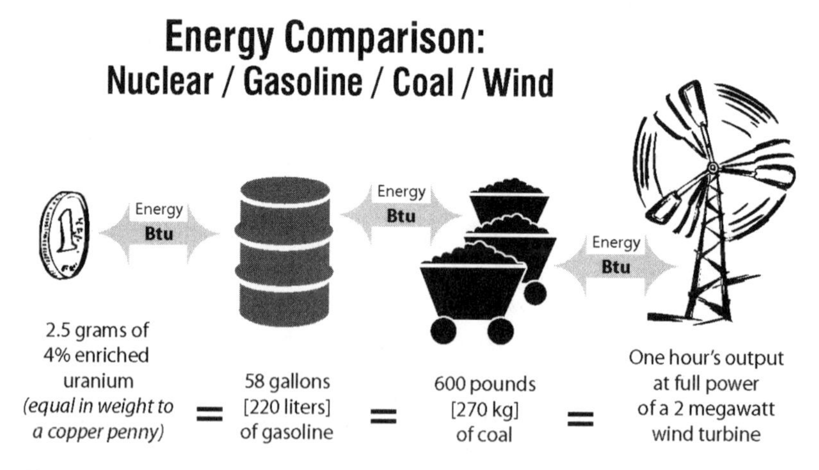

These theoretical values do not account for losses in production and transmission. The following values were use: 1 gallon of gasoline = 6 pounds; gasoline = 125,000 Btu/gallon; Coal = 12,000 Btu/pound; 1 Btu = 0.0002928 kWh

squeezed out of the picture, and continue building until all of our electricity and hydrogen generation is achieved by nuclear fission.

But this isn't going to happen. If nuclear energy gets any chance at all—which seems rather doubtful—it won't be given the starring role, or even, in the grand scheme of things, a lengthy one. Nothing lives forever. Nuclear power plants generally have lifetimes of 25 to 40 years. Any nuclear plant built today would be decommissioned within 50 years. Hopefully, this would allow us enough time to develop and deploy the new solar and wind technologies lining up on the horizon. Or perhaps even time to perfect the *fusion* reactor—utilizing a technology that turns hydrogen into helium and employs none of the heavy radioisotopes that make *fission* reactors so objectionable.

Will there be any nuclear in the hydrogen future? It's just a matter of who's willing to dance with the devil for the next half century.

But enough speculation. Now for some facts.

U.S. nuclear facilities

In the year 2002 there were 104 nuclear facilities operating in the United States. This is down from an all-time high of 111 nuclear plants in 1992 (*www.eia. doe.gov/neic/infosheets/nuclear.htm*). Despite the drop in plant numbers, electricity generation from nuclear plants in the U.S. was at its highest point to date in 2002; just over 780 billion kilowatt hours.

When burned, one pound *[.45 kg]* of coal will produce just a little over 1 kilowatt hour of electricity, and right around 2 pounds *[.9 kg]* of carbon dioxide (*http://www.seen.org/pages/db/method.shtml*). At that rate (2 lbs. of CO_2 per each kWh), nuclear power plants in the U.S. in 2002 prevented the release of over 1.5 trillion pounds *[680,400,000 kg]* of CO_2 into the atmosphere.

radioactive waste

Was it worth it? This is a rhetorical question in disguise. Despite the admirable safety record of the U.S. nuclear industry for the past 25-plus years, it's an enterprise that generates several tons of highly-lethal radioactive waste every year, and there is no general agreement on what to do with it (see below). Until that issue is resolved, nuclear energy is likely to remain mired in a bog of skepticism, if not outright contempt.

The problem lies not with the leftover uranium-238, but with the lighter elements produced from the fission of uranium-235. Generally, U-235 splits into one large atom (atomic weight around 137) and one smaller atom (atomic weight around 95). Some of these elements can be highly radioactive. Strontium-90 and cesium-137—two common fission byproducts—both have half-lives of around 30 years, which means that in any given sample, half of the atoms will decay into lighter elements in the course of 30 years. Busy little nuclei, in other words. To make matters worse, both elements can be taken up by living organisms which mistake them for more benign substances: strontium-90 mimics calcium, while cesium-137 appears wickedly close to potassium. For more about the products of U-235 fission, visit: *http://hyperphysics.phy-astr.gsu.edu/hbase/nucene/fisfrag.html#c4.*

Pebble-Bed Modular Reactors

Might there be a better way to handle the spent fuel problem? The fuel pebbles used in the pebble-bed reactor Zed spoke of—the little cannonballs—offer distinct advantages over conventional fuel pellets, in that the silicon carbide coating around the fuel pebbles is formulated to isolate all the radioisotopes for the hundreds of millennia it will take for them to decay into innocuous elements.

Of course, this does not mean that the products of radioactive decay—mostly neutrons and beta particles—are contained; only the substances that emit them. What's the difference? Imagine this: on the other side of the hill from where you live there is an escape-proof prison where there are no guards. This prison is analogous to a fuel pebble. Inside the prison many of the prisoners have guns, which they enjoy firing at random intervals. This particular prison is unique, in that bullets can pass through the walls, though the prisoners, themselves, cannot.

The gun-toting prisoners are lunatics, you see, so they don't fire at anything in particular. They just like to pull the trigger. Sometimes they shoot each other—quite by chance—and sometimes the bullets fly over and through the prison walls. If you walk too close to the prison you stand a good chance of getting hit by a stray bullet, but, since the prisoners cannot escape, you'll never have to worry about them getting out and ravaging the countryside. Sooner or later, all the prisoners will kill each other or die of old age, and the firing will cease, but since there are so many of them shooting wildly in all directions, it's going to take a long time. So the best you can do is to avoid the prison.

And, of course, hope no one breaks it open or picks it up and moves it.

For two opposing perspectives on the Pebble Bed Modular Reactor, visit *www.nei.org/doc.asp?catnum=4&catid=340&docid=&format* and *www.ieer.org/comments/energy/chny-pbr.html.*

How might the Pebble-Bed Modular Reactor best be used to make hydrogen? In a 2002 paper titled, *Production of Hydrogen By Nuclear Energy: The Enabling Technology of the Hydrogen Economy*, K. R. Schultz and others from

General Atomics summarize the results of a study they conducted for the Department of Energy. After dismissing the electrolysis of water as too inefficient (only about 36 percent efficiency overall, the researchers concluded, after taking into account the energy losses in the heat-to-electricity process) the team examined a number of possible methods and reactor designs to split water using the heat created during the transmutation of elements. The researchers concluded the best way to make hydrogen with nuclear power was to split water in a "Sulfur-Iodine Thermochemical Water-Splitting Cycle." This is a three step process that uses sulfur and iodine (all of which is recovered and reused) to split water at temperatures far below those needed for direct thermal water splitting (850 degrees Celsius in the hottest part of the reaction, compared to 2,500 degrees Celsius required for direct thermal water splitting). The researchers were confident their process would yield a 50 percent heat-to-hydrogen efficiency, using a high-temperature helium gas-cooled reactor. The pebble-bed reactor is one incarnation of that class of reactor. This paper can be downloaded at the General Atomics website: *http://web.gat.com/pubs-ext/MISCONF02/A24178.pdf.*

adaptations to solar energy

Could this technology be adapted to solar energy? Solar concentrators are capable of delivering higher temperatures than nuclear reactors (greater than 2,000 degrees Celsius), which could lead to higher system efficiencies, if not for the nagging fact that such temperatures are possible for only a few hours each day, leaving a lot of rather expensive equipment idle most of the time. What is needed, then, is a way to store the sun's energy during times when it is available in excess, and use it during the nighttime hours. Is this possible? Perhaps. See Chapter 14, *Sunlight to Watts: A Towerful Idea*, for a promising new sunlight-storing technology. In the meantime, a General Atomics presentation comparing nuclear and solar energy for hydrogen production utilizing the sulfur-iodine cycle can be found at: *http://gcep.stanford.edu/pdfs/hydrogen_workshop/Schultz.pdf.*

disasters

Zed's description of the unfortunate events at Three Mile Island in March 1979 was a brief simplification of an incredibly complicated chain of events. For a more in-depth, yet readable, analysis of what really happened, visit: *www.nucleartourist.com/events/tmi.htm.*

The Chernobyl disaster in April 1986 was a case of bad reactor design, *plus* foolhardy operational procedures, *plus* a lack of adequate safety features. An old car with no brakes, rolling down a hill toward a brick wall, in other words. For a brief, digestible account of what happened, visit the HyperPhysics website at: *http://hyperphysics.phy-astr.gsu.edu/hbase/nucene/cherno.html*.

nuclear waste

And finally: the question of nuclear power can hardly be discussed without also addressing the subject of nuclear waste. There are currently over 35,000 tons of radioactive wastes stored in numerous locations across the United States, with no general agreement on what to do with it. The federal government hopes to transport it all to Yucca Mountain, a federally-owned facility located 90 miles northwest of Las Vegas, Nevada. The feds say their repository—located in a hollowed-out mountain 1,000 feet above groundwater—is as safe as it gets, and their methods of transportation and containment (by truck and rail through 43 states, in heavily shielded containers) are sound.

The government's optimism, however, is not universally shared. The State of Nevada has filed numerous lawsuits designed to stop the site from opening in 2010, and the City of Las Vegas has vowed that no transport carrying radioactive wastes will be allowed within the city limits. Various environmental groups oppose opening the site for a plethora of reasons, ranging from geologic instability to possible contamination of groundwater, and serious questions remain concerning the vulnerability of transports to terrorist attacks.

General information on Yucca Mountain can be found on the EPA website at: *http://www.epa.gov/radiation/yucca/about.htm*. A dissenting view of the site's viability is posted on the Institute for Energy and Environmental Research website at: *http://www.ieer.org/fctsheet/yuccaalt.html*.

There...I certainly hope this has brought some small degree of clarity to a woefully murky subject.

chapter 13

The Wind as Fuel: Watts Possible?

Always the showman, I thought, catching a glimpse of Zed as I stepped out the door of my cottage the next morning. It was another beautiful fall day at the Wasserstoff Farm. I'd slept well—at least until I began dreaming about being sucked into a giant fan at Wasserstoff World—and again there had been two real eggs on my plate. Add to that three strips of not-half-bad ersatz bacon and the usual fare of fruit, toast and jelly, and Zed's excellent coffee, and it was a superb breakfast. And it had been over 24 hours since the old Druid had done anything to make gray hairs sprout like so many weeds on my head. I was ready for anything.

So, when I saw Zed riding a bright-purple mountain

bicycle with a beanie-mounted propeller on his head, it seemed like it might be a fairly laid-back day. Zed was just having fun, right? As usual, it was a simplistic conclusion.

I watched him maneuver around in tight circles and figure-eights, like a kid who'd finally graduated from training wheels and felt like showing off. But it wasn't an ordinary bicycle nor, for that matter, an ordinary propeller. And, as he worked his way closer, I saw that the beanie was really a helmet. With tubes and wires going in and out of it.

The propeller—maybe 10 inches in diameter—was attached to the shaft of a small generator, and the electrical lead coming from the generator fed into a small box just behind it, before disappearing into the helmet. On the bicycle's frame there were tubes connecting what was obviously a water storage tank and a noisy unit that looked and sounded like a pedal-driven compressor. The compressor—with its incessant *phewt, phewt*—was attached to a square, layered apparatus that must have been a fuel cell. A pair of wires from the fuel cell fed into an electric motor. The motor itself was tied into the bicycle's chain-drive mechanism, rounding out the strange medley of gadgets and parts.

I had to laugh. Zed resembled a modern-day Frankenstein, using wind instead of lightning to bring himself to life.

Upon hearing me chuckle, he spun his bike around and brought it to a screeching halt right in front of me. "What's so blasted funny?" he demanded.

I noticed his T-shirt for the first time: *Watts Up With Wind?*, it read. "Why don't you go look in a mirror?" I suggested, with a smirk.

"Go look in one, yourself, sonny boy. I'm busy," came the terse reply.

I could see that poking fun, however lightly, was getting me nowhere. "Sorry, Zed. It's just that you look like..."

"Look like *what?*" he snarled.

"Uh...you look like you've got something important to show me here."

With the slightest trace of a smile, he said, "Nice save, lad. Just when I thought you were about to crash and burn."

Trying to change the subject, I asked, "So what do you have here, anyway, Zed?"

"What else? It's a wind-powered, hydrogen-fueled electric bicycle."

"Get out! No way can you harness enough wind to drive that thing,"

I said, feeling very fortunate he hadn't tried to do this with U-235.

"It's not a production model," he answered, a bit defensively, "but it *does* work." He spun the bike around, pedaled a little ways to get his balance, then flipped a switch on the handlebars. To my surprise, it took off like a well-tuned motocross bike. Zed had to throw himself forward to keep the front wheel on the ground. He tore across the meadow for a couple hundred yards, slammed his foot to the ground, sharply pivoting around a rock and then made a beeline back to where I stood.

"That's pretty cool, Zed! But how far do you have to pedal the thing to make enough hydrogen to do that?"

"Into a really stiff wind? About twelve miles."

"Impressive."

"Humph," he groaned. "The whole purpose of this wonder of backyard engineering was to demonstrate a concept. But I can see now it's over your head."

If he was trying to make me feel bad, I had to admit it was working. "Sorry, Zed. I really do think it's neat. Care to explain how it works?"

He set the kickstand with his heel, slung the helmet over the seat, and walked away. "Not really," he said, the words coming from over his shoulder.

I swear—there's nothing worse than a touchy wizard. I ran up behind him, pleading, "C'mon, Zed! I was just playing around; I *really, really, really* want to know how it works."

He spun on his heels, saying, "Oh, all right. But I'm only going through this once, so pay attention."

"Sure, Zed," I answered, "anything you say."

Returning to the bike, he picked up the helmet and gave the two-blade propeller a spin with his finger. "First, we have to determine how much power is available. To do this we have to start with the propeller." He glanced up to see if I was paying attention, then said, "It has a diameter of 0.25 meters, or a little under 10 inches. By taking pi times the square of the radius—0.125 meters—we determine that it has a sweep area of roughly 0.05 square meters. *Comprende?*"

"I'm with you."

"Good. Now, the power of the wind does not increase arithmetically, as you might think; it increases with the cube of the wind speed. A 30 mph wind blows with over three times the force of a 20 mph wind, so a little extra wind can make a big difference. For the sake of this demonstration I'll assume a 30 mph wind. You can achieve this by riding 30 mph on a calm day, or pedaling 10 mph into a 20 mph wind, or whatever. Or you can strap the helmet down facing into a 30 mph wind and go take a nap."

"So how much power can it produce in a 30 mph wind?" I asked.

"I was just getting to that. The force—or power density—of the wind is 0.05472 times the wind speed cubed. So 30 times 30 times 30 is...27,000. Multipy that times 0.05472 and you get about...1,477.

"1,477 what?"

"Oh, right...watts per square meter. So if our propeller is 0.05 meters, and we multiply that by 1,477 watts per square meter, we see that the energy of the wind hitting it at 30 mph would be about 74 watts."

"Not bad, Zed."

He rubbed his chin and said, "No, but the propeller can't possibly convert all that wind into usable energy."

"And why not?"

"Think about it. The only way a propeller could capture all the wind's energy would be to stop the wind. But if it did that, the propeller wouldn't turn, now would it? In fact, there's a mathematical proof called the Betz limit. It sets the theoretical maximum efficiency of a wind turbine at 59.3 percent."

"Oh...well, that's not so bad."

"True, but even that's pie in the sky. The Betz limit is the unreachable ideal. It's like the perfect circle, or the selfless mass-tort lawyer. It doesn't actually correspond to anything in the real world."

I narrowed my eyes, and said, "So, what are you telling me?"

Zed said, "Simple. If you can get 30 percent, take it and run."

"But 30 percent is only..."

"...Only 22 watts, out of 74, I'm afraid."

"That stinks," I grumbled.

Zed disagreed. "No it doesn't," he said, shaking his head. "If I had lied to you and told you that the power density of a 30 mph wind upon the bicycle's tiny turbine was 25 watts, you'd be doing handsprings at the prospect of netting 22 watts."

"Yeah, but..."

"Look at it this way: what do you have to do to get that 22 watts of power? Nothing! You don't have to dig up any coal, or heat any chemicals, or make steam, or burn one drop of gasoline. All you have to do is stick the propeller into the wind to reap the bounty. It's like raising vegetables. Do you feel cheated because you can't eat radish leaves? No, you don't. But that's still a lot of food energy going to waste, isn't it?"

I was beginning to see his point. Nature was rife with inefficient energy conversions, but she hardly suffered for it. Just because a leaf only converts a paltry 3 percent of the available sunlight into useful energy doesn't mean there's a dearth of greenery on the planet. Three percent is all the plant needs—any more and things would quickly get out of balance.

I ask Zed why this was so, and he replied, "Simple. In her slow and plodding way, Mother Nature generally gets things right the first time, while we have this regrettable habit of screwing things up and not admitting it until it's almost too late. Then we have to redouble our efforts in a different direction just to keep our heads above water. In a sense, human history is just a long chronology of wasted effort."

"So you're saying I should be happy with 22 watts?" I asked.

"If you're smart, you will be."

"Okay, I'm happy. Now what happens?"

"Well, we lose another 5 percent when the turbine's 3-phase alternating current is converted to direct current to run the electrolyzer."

"I beg your pardon?" I was feeling stupid. Again.

Zed sez

Today's cars are only about 20 percent efficient at converting the chemical energy in gasoline into forward motion. But that gasoline doesn't just blow in on the wind. First it has to be pulled out of the ground as crude oil from a mile or more below the surface, then transported—usually across an ocean or two—before being sent to the refinery, where it's chemically altered, stored, and later transported again. If you can find anything efficient about that, you just let me know.

Zed smiled at my confusion. He said, "A wind turbine is really just three pairs of magnets of opposite polarity, evenly-spaced around the inside of a drum, or a rotor. When the magnets turn around a trio of copper wire windings, called the stator, they induce an electrical current in the wires. Since the magnets are of opposite polarity, every time the rotor turns a revolution the current in each of the three windings changes from positive to negative, and back again, and they do it in phase. Graphically, it's called a sinusoidal wave, or simply a sine wave..." He paused, and asked, "Are you still with me, lad?"

"After a fashion," I answered.

"Good. The point is, with three pairs of magnets and three sets of copper windings, you are producing three 'in-phase' sine waves of alternating current. And the electrolyzer wouldn't have the slightest idea what to do with it. All it understands is direct current, which is just a stream of electrons traveling from a higher electrical potential to a lower one, like what comes out of a battery."

"How do you rectify *that* situation?" I asked.

"With a rectifier," he answered.

"You're kidding."

"Not at all. That's what it's called. Or they. It takes three—one for each phase. By the time it's all said and done, you have a single circuit of direct current at a constant voltage. But it costs you 5 percent, or about 1 watt. So we're down to 21 watts."

I said, "We can afford it. What's the next hit?"

"The electrolyzer. We'll be realistic and give it an efficiency of around 75 percent. So the hydrogen produced by the electrolyzer has an energy value of around 16 watts of the original 74 watts present in the 30 mph wind."

We were down to 21 percent total efficiency and hadn't even gotten to the fuel cell or the electric motor. Overlooking the compressor, which seemed to get its energy from the pedals, I asked Zed, "And what's the efficiency of the fuel cell?"

"Along with the motor, about 50 percent, give or take."

"So we're down to 8 watts of power delivered to the wheels for every 74 watts of wind energy hitting the propeller? That's just a little more than 10 percent. I know we talked about this earlier, Zed, but really...10 percent?"

"Yeah, it's pretty amazing, alright."

I raised an eyebrow over his choice of words. "Amazing?" I repeated, incredulously. "Don't you mean 'abysmal'?"

He speared me with that cunning, predatory look that meant I'd just walked into a trap. It didn't take long for the jaws to close. "No, I always mean what I say, lad. Because it is amazing that you can take the wind's raw kinetic energy and convert it into energy from a chain that begins with mechanical force, goes to alternating current, then direct current to hydrogen fuel, then back to direct current and mechanical force. And still have any efficiency at all.

"Contrast that with a gasoline-powered automobile. On average, today's cars are only about 20 percent efficient at converting the chemical energy in gasoline into forward motion. But that gasoline doesn't just blow in on the wind. First it has to be pulled out of the ground as crude oil from a mile or more below the surface, then transported—usually across an ocean or two—before being sent to the refinery, where it's chemically altered, stored, and later transported again. If you can find anything efficient about that, you just let me know, okay?"

Touché, Zed.

"I see your point," I admitted, "but it seems like a tremendous amount of pedaling into the wind just to build up enough hydrogen to go zooming a few hundred yards across a meadow."

Zed pointed to his component-laden purple bike and said, "This thing? Get serious! It's like crossing a falcon with an ostrich and hoping to get something that flies. No one in their right mind would build something like..."

"*You* did," I interrupted.

"Hmmmm? Oh, yes, so I did. And a good thing, too. But it's only a teaching tool, you understand. If you really wanted a hydrogen-cycle you'd mount a big windmill on a high tower, not a teensy one on a helmet. Then you'd simply transfer the hydrogen from the storage tank by the tower, into the bicycle. Surely you don't think...?"

"Of course not, Zed."

"Good, glad to hear it. But enough of this small-time foolishness. Now that you understand the basics, let's go see a *real* wind turbine."

I looked around, but didn't see one. "Where is it, Zed?"

"Close, really close. Just follow me," he said, somewhat evasively, I thought. Before I could answer, he walked away in the direction of the dimensional compressor.

"Wait a minute, Zed," I objected, "I don't know if I want..." The shed that housed the spooky thing was several yards away, but the door was wide open and I could see it leering at me. I was already starting to feel funny. Tingly, inside and out. Like I'd just swallowed a pot of radioactive coffee. The air was saturated with the acrid odor of ozone. Before me Zed—or at least a blurry version of him—motioned for me to follow. I turned and looked behind me. There was nothing but impenetrable fog. Reluctantly, I walked forward.

When I caught up with him in the midst of the murk he grabbed me by the arm and said, "Okay, here we go." Before I could object, we were whisked into a dimensional maelstrom. It was like being the central point around which the rest of the world quickly revolved. I closed my eyes and held onto Zed since he, unlike me, knew where we were going. In a moment the storm grew quiet and I could see that we had been transported to the inside of a strange structure where an assortment of gear boxes and shafts loomed menacingly.

"So, what do you think?" Zed asked, once my head stopped spinning. "I stayed up most of the night working it over. I expanded the depth of the electro-dimensional field, so it's not such a shock to the system."

"I can see that. Now it's more like pulling a piece of tape off my skin slowly, instead of all at once," I answered truthfully.

He shrugged. "Maybe so. But at least my hair stays down."

"It's good to have priorities," I agreed.

"Anyway," he said, spreading his arms expansively, "here we are. Inside the nacelle of a 3-megawatt commercial wind turbine. A Pickett Meg-3, to be exact."

A Pickett Meg-3? Oh, brother. I held my tongue and asked, "What's a 'nacelle'?"

"Like an enclosure. But with one less syllable."

"Oh," I said.

Looking around for the first time, I was shocked by the sheer scale of the thing. The distance from one end to the other was at least 25 or 30 feet, and we could both stand up inside without difficulty. It was gratifyingly bright inside, but noisy. I felt like an ant inside of a well-lit transmission.

"It's a little complicated, so we'll just hit the high spots. You see, right here we have the main shaft, coming from the propeller blades." He was pointing to an enormous shaft turning within a proportionately large bearing.

"Why is it spinning so slowly?" I wondered.

"That's just the way it's designed. Even at the optimal rate of 20 revolutions per minute, the tips of the blades are traveling at over 200 miles per hour. If it spun as fast as a small wind turbine it would tear itself apart in a heartbeat."

That made sense, but I was confused. I asked, "How can it generate electricity, if it's turning so slowly?"

He pointed to a heavy steel box to his left, into which the main shaft disappeared. It made a high-pitched whine that my ears took an instant dislike to. "This is the gearbox. In here the 20 rpm is kicked up to 1,500 rpm. From the gearbox a smaller shaft feeds into the generator housing, here." To the left of the gearbox, toward the back of the nacelle, a feverishly turning shaft vanished into another heavily-constructed steel casing. "The magnets and windings are all in here. The current coming out is 690 volts, 3-phase AC."

I recalled the explanation Zed had given earlier about the components of his purple "wind-powered, hydrogen-fueled electric bicycle," and asked, "Where's the rest of it, Zed? This unit *is* for hydrogen production, right?"

"Of course it is. We *are* at Wasserstoff World, after all."

"Back there again, huh? Glad you cleared that up. Half the time I don't know where we are."

"And that's *my* fault?" he bristled, shaking his hairy head. "It might help if you'd pay attention." He shuffled past me and stood on a small platform near the center of the enclo...uh, the nacelle. "Stand here, close to me," he ordered. I did. Immediately the platform began to sink into a brightly-lit Plexiglas tube. "This is the inside of the tower," he explained, as the platform descended. "It may look empty, but it's really filled with hydrogen gas, compressed to 10 atmospheres."

In spite of the lights along the inside of the tower, it was still eerily tomb-like; like being inside a gutted submarine, or perhaps an oversized grain silo.

The tower was huge, and grew steadily larger in diameter as we went down. And down, and down. Suspicious, I asked, "Zed, did you shrink us, like you did in the yellow submarine, or is this thing really this big?"

He answered, "No, lad. No tricks here. Everything—including you—is to full-scale. From the nacelle to the base of the tower is nearly 300 feet, or the height of a 30 story building."

Presently the elevator dropped into a giant, circular building at the base of the tower. I stepped off the platform and looked around. "Where's the electrolyzer?" I asked.

Zed answered, "This is just a wild guess, but I'm betting it's that big, grey unit with 'Electrolyzer' painted on the side."

"Uh, sure...okay."

"Next to that is the compressor for compressing the hydrogen gas. Beyond that are the batteries, the power and electronics interface, and the fuel-cell stack."

"Batteries? Fuel cells? What for?"

"Big, commercial electrolyzers like this one aren't things you turn on and off like a lightbulb, lad. They should be maintained at some degree of operation even when the wind doesn't blow, and that's the role of the batteries. The wind charges the batteries, but if the wind doesn't blow, the fuel cells step in to charge them. Likewise, the fuel cells provide power to the electrolyzer and everything else in this room when the turbine is inactive."

Something wasn't adding up. I asked, "What kind of current does the electrolyzer take?"

Zed answered, "Direct current, around 1.25 volts."

"So this 3 million watts of 3-phase, 690-volt alternating current is all stepped down to 1.25-volt DC?"

Zed laughed. "You do have a gift for confusing yourself, don't you, lad? No...most of the power is stepped up to several thousand volts and sent to the electrical grid. Hydrogen production is more of an avocation—something that's done when there's a surplus of wind, or a drop in electrical demand, like at night. The idea is to design the wind farm with the potential to produce more power than it's contracted to supply. In that way, the hydrogen is like icing on the cake. And by doing the electrolysis onsite, with only one AC to DC conversion, the process is that much more efficient."

He opened a door at the base of the tower, and we stepped out into a bright, breezy day. Looking up, first at the nacelle 30 stories up, then at the three 150-foot propeller blades, sweeping out an area nearly equal to 1.5 acres, I felt absolutely Lilliputian. Zed gestured toward the blades, which sounded like the wings of a dozen whale-sized birds, and said, "For an extra buck-fifty I can arrange a ride on the tip of one of the blades. It makes a Ferris wheel seem like a slow-motion carousel."

I just smiled at the old wizard, and said, "My cash is in my saddlebags. Do you take plastic?"

A BONUS 1.3 MW turbine at the Uhre Wind Farm in Denmark. *Photo © BONUS Energy A/S. www.bonus.dk*

technistoff - 13

Yes, commerical wind turbines really are that big.

For anyone planning to build a wind-powered, hydrogen-fueled electric bicycle like Zed's, you should know that the color will be the easiest thing to duplicate. At least two of the main components are proprietary. So much so, in fact, that no one else on the planet has access to them. I'm speaking, of course, of a fist-sized electric motor capable of producing enough horsepower and torque to accelerate a cycle and its rider like a finely-tuned dirt bike. Or the super-compact fuel cell that provided the quick surge of electricity from hydrogen. If you suspect wizardry, in addition to—or in lieu of—savvy engineering, you are not alone.

Zed's numbers are, however, quite believable. There really is a theoretical limit to the amount of energy you can extract from wind—59.3 percent of it—and if you are interested in learning about the Betz limit and related phenomena, visit: *www.greeleynet.com/~cmorrison/windcalc.html*. The tables found at this website are calculated from formulas provided in Paul Gipe's excellent book, *Wind Power for Home and Business*, as is the power density formula given below.

efficiency & wind power

A word about efficiency: when speaking about wind power, the word "efficiency" simply loses all the daunting implications that it bears in discussions about, say, coal-fired power plants. When you are digging thousands of tons of coal out of the ground and transporting it hundreds of miles to a town-sized power plant, the difference between 40 and 45 percent efficiency—in terms of power produced, and emissions avoided—can be staggering.

With wind turbines, efficiency is a far more relative, and benign, concept.

The 30 percent efficiency Zed attributed to his tiny 10-inch wind turbine was probably optimistic. Most small wind generators fall into the 12 to 15 percent range, while the large commercial turbines are closer to 20 percent. The important thing to remember is this: it's not that important. The area swept by a propeller increases in proportion to the square of the radius, so if Zed had made his little propeller 12 inches *[30.5 cm]* in diameter (instead of 10 inches *[25.4 cm]*) and claimed a mere 20 percent efficiency, he still would have come out acceptably close to 22 watts in a 30 mph *[13.4 m/sec]* wind.

Far more important than the area swept by the propeller is the average speed of the wind where the turbine will be placed. That's because—as Zed noted—the force of the wind increases as a function of the *cube* of its speed:

Wind Power Density (watts/m^2) = .05472 x wind speed (mph^3)
For those who speak metric: **= .6127 x wind speed (m/sec^3)**

Thus, a 30 mph wind packs 3.38 times the punch of a 20 mph wind. A hurricane, with wind speeds of 150 mph, is pushing with the force of 185 kilowatts per square meter—125 times more forceful than a stiff 30 mph breeze.

Why does the number 30 mph keep popping up in this discussion? It just works out to be very close to the highest wind speed most turbines can operate at and remain stable. For winds much beyond that speed, turbine manufacturers, large and small, build safeguards into their turbines to protect them from spinning out of control or otherwise sustaining damage.

Every scheme I am aware of that uses wind to produce hydrogen does so by the electrolysis of water. The system Zed had set up at Wasserstoff World was similar to one suggested by L. J. Fingersh in *Optimized Hydrogen and Electricity Generation from Wind*, a June 2003 report produced for NREL, and available at: *http://www.nrel.gov/docs/fy03osti/34364.pdf.*

using wind for hydrogen

Fingersh's idea is simple and elegant. Essentially, his plan is this: propose a wind farm to supply a preset amount of power to the electrical grid, then engi-

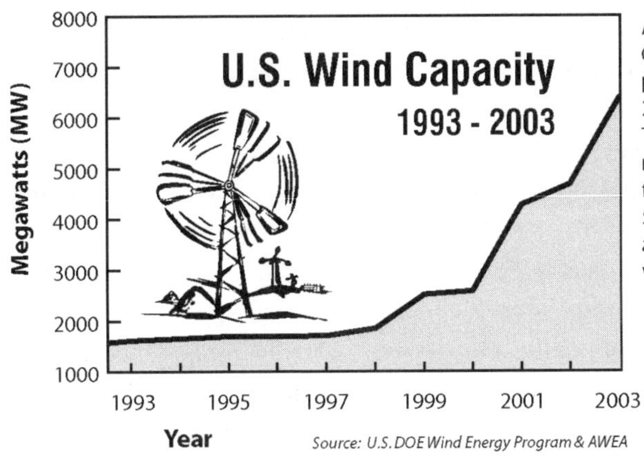

As of January 2004, California led the pack with over 2,042 MW installed. Texas is next with nearly 1,300 MW; then Minnesota at 562 MW and Iowa at 471 MW of wind energy.

neer it in such a way that it will produce an excess. Promise 500 kilowatts to the grid, for instance, then design the system to produce 750 kilowatts, with the extra power going into hydrogen production. By adjusting grid commitments and wind speed he offers several different scenarios, ranging from very little power going into hydrogen production, to more than half to hydrogen. The amount of hydrogen produced would vary, according to local conditions, as Fingersh explains:

> *"For example, one might choose to provide more electricity and less hydrogen if the winds are steady and grid needs are high (as in California). One might also choose to produce more hydrogen and less electricity in locations with strong winds but small electrical loads (as in North Dakota)."*

Two key aspects of his plan deserve attention. First, he suggests that the electrolyzer be located onsite. In this way, the low-voltage direct current needed by the electrolyzer can be supplied through the turbine's power electronics. The AC input on the electrolyzer could be eliminated, so the AC to DC conversion (within the electrolyzer) could be avoided. Since every conversion comes with a price tag (from 3 to 5 percent efficiency, according to Fingersh), such a system would optimize hydrogen production.

Secondly, he proposes using the tower itself (with a few modifications) for hydrogen storage, thereby eliminating the need for an external tank. Using a 1.5 megawatt turbine as an example (with tower heights varying from around 210 to 280 feet *[64 to 85 meters]*), Fingersh calculates that 13.3 to 17.3 megawatt hours of hydrogen could be stored inside each tower at relatively low pressure. Although this is less than 800 pounds *[263 kg]* of hydrogen per tower, it would be ample storage, since hydrogen would routinely be piped away and replenished.

As noted by Zed, while standing inside his giant *Pickett Meg-3* wind turbine at Wasserstoff World, the limiting factor in any system using wind to produce hydrogen is the electrolyzer. They are finicky devices that can only achieve optimum efficiency within a well-defined range of voltage, amperage, temperature and pressure. They most certainly do not like to be turned on and off.

wind electrolyzer

Any wind/hydrogen system needs, first and foremost, to address the needs of the electrolyzer. The electronic configuration must be such that the varying AC voltage and amperage coming from the turbine can be continually converted into the steady DC power the electrolyzer needs. The fuel cell and battery backup system employed to run the electrolyzer and related hardware during calm times must also supply power within the electrolyzer's limited parameters. Though it may prove to be a complex task, it is probably more easily done in a wind farm where at least a portion of the power is dedicated to the grid. The electrolyzer and the grid can serve as buffers for one another; power not needed by the electrolyzer can be shunted to the grid, just as the extra power produced during particularly windy times can be directed to the electrolyzer.

In a stand-alone system, where the options to redirect excess energy are severely limited, the problems are compounded. Ideally, the electrolyzer should be sized to take full advantage of the available energy during windy times, without it being so large that it would quickly gobble up energy reserves during calm periods. A paper that discusses this and other problems, *A Study of a Stand-alone Wind and Hydrogen System*, was prepared by Lars Nesje Grimsmo, and others, from The Norwegian University of Science and Technology, and presented at the March 2004 Nordic Wind Power Conference at Chalmers University of Technology (*www.elkraft.chalmers.se/Publikationer/EMKE.publ/*

Middelgrunden Offshore, Denmark. *Photo © BONUS Energy A/S. www.bonus.dk*

NWPC04/papers/GRIMSMO.PDF). The researchers conclude, as I read between the lines, that there is no simple solution to sizing all the individual components (turbine, electrolyzer, storage tank, compressor, etc.) in a stand-alone wind/hydrogen system but that their models are accurate enough to proceed, until the knowledge gained by experience can better direct the process.

Putting the issue of hydrogen production aside for the moment, how do wind farms stack up against coal-fired power plants in terms of sheer output? In 2002, coal-fired power plants operated by electrical utilities in the United States produced over 1.5 million gigawatt hours of electricity in 1,566 electrical generators (*www.eia.doe.gov/cneaf/electricity/epa/epa_sum.html*) for an average of around 958 gigawatt hours per generator (most facilities have multiple generators).

very large wind farms

In contrast, a *very large* wind farm, such as the Eurowind facility currently being built offshore near Lillgrund, between Denmark and Sweden, will employ 48 wind turbines, rated at 1.8 megawatts each, to produce a combined annual output of 300 gigawatt hours of electricity (*www.eurowind.se/eng/sida_lillgrund.html*), or 31 percent of the power of a single coal-fired generator.

There *are* bigger wind farms. The Horns Rev project mentioned below is

The Horns Rev Offshore Wind Farm situated far out at sea has developed new concepts for wind turbine maintenance, including the possibility that service personnel can be landed at the nacelle from a helicopter. *Photo © Copyright: Elsam A/S. www.hornsrev.dk*

estimated to produce 600 gigawatt hours of electricity per year. The U.S. would need to install 2,500 such wind farms to produce the electricity currently being produced by the burning of coal.

Clearly, wind power has a long way to go before it's able to squeeze coal out of the picture, but the future is, nonetheless, bright. The U.S. wind industry has grown from a mere 10 megawatts in 1981 to 6,374 megawatts in 2003, with nearly 60 percent of that growth coming since the beginning of the millennium when, in the year 2000, the total installed U.S. capacity was at 2,578 megawatts. These figures are available on the American Wind Energy Association website: *www.awea.org/faq/instcap.html*.

New Wind Power Installed Worldwide in 2003

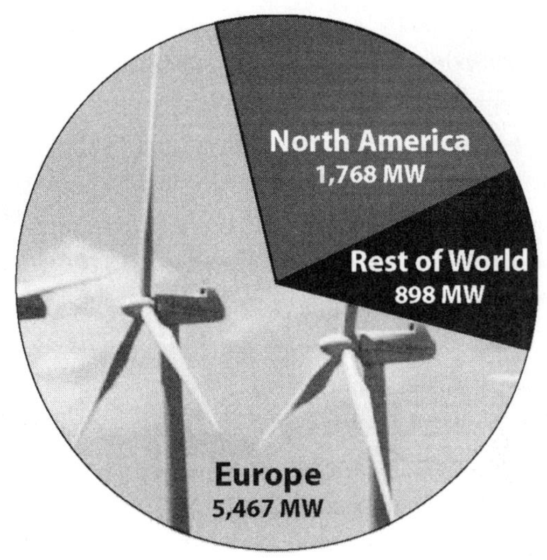

MW = Megawatts
Source: EWEA / AWEA

world-wide wind power

Though it appears that Americans are finally taking wind energy seriously, the U.S. efforts pale in comparison to Europe, who leads the way in wind power with 28,706 megawatts of capacity installed at the end of 2003, giving Europe 73 percent of the world's wind energy pie. These figures, along with many others, are available at the European Wind Energy Association website: *www.ewea.org*.

Wind technology is at last coming of age, with several large manufacturers diligently working to keep the field competitive and innovative. As more and more people express a willingness to pay a little extra for wind generated electricity—and coal's reputation becomes proportionately more sullied—the price of wind energy will drop, as the prevalence of wind farms increases. Since 1975—when the price of wind-generated electricity was about $0.30 per kWh—the price has continued to fall. In many areas of the U.S. and Europe, electricity from wind is now cheaper than electricity from coal. As wind power becomes more prevalent and the hydrogen economy emerges from the laboratory, we should expect to see wind-produced hydrogen competing with hydrogen extracted from fossil fuels.

Flying Electric Generators

Ever jump on an airplane on a perfectly calm day, only to see your knuckles turn white from gripping the armrests as the plane rocks and tosses on its way up to cruising altitude? The fact is, there's a lot more wind up high than there is down here, and Sky Windpower Corporation, based in Ramona, California, is hoping to tap into it. Their concept is as bold as it is simple: station an electricity-generating rotocraft at 15,000 feet or so, then tether it to the ground with its own (very special) electrical cable. When the high winds blow—which is better than 90 percent of the time in some prime locations—the FEGs (Flying Electric Generators) deliver power to a ground-based power station. And when the winds aren't cooperative? The craft is kept aloft by electricity supplied from below. Sky Windpower's feasibility studies suggest that, done on a large enough scale, their flying generators could produce electricity for under $0.02 per kilowatt hour, and could therefore produce hydrogen from water—via ground-based electrolyzers—cheaper than it is currently being produced from fossil fuels. It's a wild idea that just might work. For more information, visit the Sky Windpower website at: *http://skywindpower.com/ww/index.htm*.

how big are the turbines?

Okay: are the big turbines really as big as Zed led me to believe? Indeed they are. Though a little smaller than Zed's 3-megawatt behemoth, the Vestas V80 2-megawatt turbine—80 of which were erected in 2002 by Elsam Corporation at Horns Rev in the North Sea—is impressive. The rotor cuts a 262-foot *[80 m]* diameter circle out of the wind, or an area equal to 1.24 acres *[0.5 hectares]*. The rotor hub is 230 feet *[70 m]* above the water. Each of the three 131-foot *[40 m]* propeller blades weighs in at 14,300 pounds *[6,490 kg]*. Not including the foundation, each machine tips the scales at over 540,000 pounds *[245,000 kg]*.

You can easily stand up and work inside the nacelle. And there really are elevators, though they are not enclosed in cool Plexiglas tubes.

Horns Rev Offshore Wind Farm in the North Sea
Photos © Elsam A/S
www.hornsrev.dk

Left: The foundations, which are monopiles (steel pipes) with a diameter of approximately 4 meters, were rammed about 25 meters into the seabed.

Below: Wind turbine erection was made by specially built jack-up vessels which can stand firm on the seabed by means of submersible supporting legs. The vessels carried two complete turbines at a time and could lift each turbine into position on the foundation with cranes on the vessels.

To learn more about the turbines, and the Elsam Horns Rev project, visit: *www.mumm.ac.be/Common/Windmills/SPE/Bijlage/* and select "Horns" brochure or visit *www.hornsrev.dk.*

For an informative article on large wind turbines, including who builds them and what big projects are in the works, read the Renewable Energy World article, *Great Expectations: Large Wind Turbines*, at: *www.jxj.com/magsandj/rew/2001_03/great_expectations.html.*

During the summer of 2002, Elsam erected the world's largest offshore wind farm at Horns Rev in the North Sea. Comprised of 80 wind turbines (Vestas V80 - 2 megawatts each) erected 14 to 20 kilometers out at sea, it will cover almost 2 percent of the total Danish power consumption. *Photo © Elsam A/S. www.hornsrev.dk*

chapter 14

Sunlight to Watts: A Towerful Idea

I never did get my ride. As it turned out, the crane Zed wanted to use to hoist me 150 feet in the air to a blade-mounted swivel chair—which he was also unable to locate—was "somewhere else." I asked Zed why a wizard of his caliber—someone who could shrink a submarine and its occupants to the size of a hydrogen atom—would even need a crane. Not surprisingly, he was maddeningly elusive about the whole thing. I suspected that perhaps his powers of illusion might begin to diminish as they became entangled with the force of gravity. As usual I was wrong—way wrong—though it would take a couple of days to find that out.

Zed had one more attraction to show me before we left Wasserstoff World. His "Saltern Tower of Power," as he called it, which made no sense, at the time. After a long, sweaty hike to the top of a steep hill in the Martian-like countryside, we came upon a huge other-worldly facility sprawled out over several acres on the vast plain below. It looked vaguely familiar. Then I remembered what it reminded me of: the direct solar-thermal water-splitting setup back at the Wasserstoff Farm. Only bigger. A lot bigger.

"Don't you already have one of these?" I inquired, scanning the hundreds of ground-mounted mirrors that reflected sunlight to a strange apparatus mounted on top of a giant lattice tower.

Zed shook his head. "Take a closer look at the tower, lad. Do you see any difference?"

I studied the tower for a moment before I saw what Zed was talking about. Instead of a large parabolic mirror focusing concentrated sunlight onto a reactor vessel on the ground below, all the sunlight here was being absorbed by a large, cylindrical collector. Resting regally atop an inverted, truncated pyramid, the collector (or the "central receiver," as Zed called it) looked like a large brightly-glowing lampshade with overdone white ruffles around the top and bottom.

"Okay, I give up. What's it do?" I asked.

"It makes electricity from sunlight—and a lot of it," he replied.

I looked over the receiver closely. It didn't resemble any photovoltaic device I'd ever seen; nothing about it gave me the sense that a direct conversion of sunlight to electricity was taking place. As large as the receiver was, it had only a fraction the surface area of the mirrors, so—if it *was* photovoltaic—it would've been much more efficient just to nix the tower and mirrors and use all the acreage to set up a massive solar array. But if not photovoltaics, then what?

Finally, not seeing any way electricity could be produced here, I said, "You're kidding."

"Hardly," he answered, smugly crossing his arms and pointing his nose toward the sky.

"How, then?"

He replied, "You figure it out. C'mon, let's get closer."

As we trod down the hill to the plain below, I began to get an idea of just how huge this so-called Saltern Tower of Power really was. The tower was at least as tall as the tower of the Pickett Meg-3 wind turbine—didn't Zed say that was 300 feet?—and the area covered by the heliostats was truly impressive; at least 30 or 40 acres, maybe more.

At the bottom of the hill we began to weave our way through row upon curved row of heliostats, each higher than we were tall. They seemed to be moving, almost imperceptibly, like the hands of a clock. I asked Zed if they really were moving, or if it was just my imagination. He replied, "Oh, no... they're moving. Each one moves according to computerized instructions in a predetermined way, to optimize the amount of sunlight reflected onto the receiver."

Once inside the multitude of heliostats, all seeming to stare fixedly at the receiver like so many stoned groupies at a rock concert, I surveyed the grounds in hopes of making some sense of this new process for making electricity from sunlight. Most noticeable were a pair of enormous cylindrical tanks at the base of the tower, each standing over 30 feet tall and at least as wide. The two tanks were separated by a very complex piece of machinery. Multiple large-diameter pipes ran in and out of the strange machine, connecting the two large vertical tanks together through a smaller, horizontal tank beside it. From the heart of the machine rose the unmistakable sound of a high-speed turbine generator.

I was beginning to get an inkling for how it might work, but I wasn't sure. "Help me out a little here, Zed, if you don't mind."

"Have you got a quarter?" he asked, "I'm hoping to show a profit on this place, you know."

"A quarter? I really doubt it," I said, fishing in my pockets. "I told you, I left all my...hey! A quarter!"

Smiling impishly, the crafty old conjurer said, "Good, lad. Now stick it in the slot."

Slot? What slot? I looked down and saw...a slot. It was sized for a quarter and set into the top of a pedestal just to my right. I dropped the quarter in and listened for it to work its way down through the mechanism. When it hit bottom I heard a confirmatory *ka-ching*, and Zed said, "There. That ought to make things a little more clear."

I don't know what I expected, but I was disappointed to see that everything looked the same...but wait, did it? Where before everything was drab black or silver, now it was all in shades of bright blue and red. The tank to my right glowed an intense red, as did the pipes leading in and out of it. The other tank, along with its complement of pipes, was blue. A blue line leading from this second tank rose up the middle of the tower and disappeared into the center of the receiver, high overhead. From the outer part of the receiver a red pipe followed the inside edge of the tower to the first, bright-red tank. But what was happening inside the receiver?

I asked Zed, who said, "Another quarter, please."

Again I found a quarter (though I'm certain it wasn't there a moment before) and inserted it in the slot.

Suddenly, both tanks, all the pipes and the receiver, itself, vanished. All that remained was the fluid flowing through the invisible pipes and tanks. I could now see that the receiver was really just a series of interconnected pipes, like a boiler. Cold fluid entered in the middle and gradually changed from blue to red as it worked its way toward the outside layer of pipes, which glowed brightly hot.

"It must be water," I surmised, "heated in the receiver and used to run a steam generator. This is like a coal-fired plant, except that it uses sunlight as the source of heat."

"Good, lad, very good," Zed praised, "but you're missing one thing here. If all it was doing was heating water, what would be the point of these huge storage tanks?"

I walked closer and examined the fluid flowing in the invisible pipes. The pipe leading into the turbine from the horizontal tank (which, I could see, in its naked repose, was actually a heat exchanger) obviously carried steam, while the pipe coming back into the heat exchanger carried plain, transparent water.

But the fluid in the tanks and in the pipes going to and from the receiver was different. It was opaque and seemed more viscous than water; almost like flowing lava. And it glowed much hotter than the water or even the steam. Whatever it was, it was able to absorb more heat than water, and to hold that heat longer. Then it hit me: that had to be the reason for the storage tanks: to store up a surplus of the sun's heat within this strange fluid during the day, and cycle it through the heat exchanger at night. If I was right, it would be possible to use the sun's energy to make electricity 24 hours a day. What a concept!

But what sort of exotic fluid had Zed discovered to make it all work? Was it even from this planet? And was it expensive? When I asked him, he replied simply, "Fertilizer."

"Fertilizer?" I echoed, incredulously.

"I believe I said that. But yes, fertilizer. A mixture of sodium and potassium nitrates, to be exact. Chemically, they are both salts, so technically this facility is known as a 'Molten Salt Solar Central Receiver Power Plant'."

Flaunting my keen intellect, I exclaimed, "Wow, this is cool!"

Ignoring my enthusiasm, Zed replied, "Actually, the salt is quite hot. Even on the cold storage side it never falls below 290 degrees Celsius, about 70 degrees above its freezing point. And the hot salt used to produce the steam to run the turbine is a sizzling 565 degrees Celsius. That's over 238 degrees above the melting point of lead. It makes for a very efficient heat-storage medium."

"How much power does this apparatus produce?"

"This is a rather small plant, designed to produce around 5 megawatts of power. That's two more megawatts than the Pickett Meg-3 wind turbine on the other side of the hill."

I scanned the field of heliostats to get a sense for the area they occupied. It seemed about the same as a small farm. I asked Zed, who answered, "About 10 acres per megawatt, so this facility takes up around 50 acres."

"That's a lot, Zed. More than a wind farm of comparable size."

"Not much more, once you consider that wind turbines need to be spaced at least four rotor diameters apart. Besides, it's not like the two technologies compete with one another for space. Hot, cloudless places are generally not good places to reap a lot of wind energy."

"This place being the exception," I pointed out.

"Uh, yes. That it is."

"And just where *is* this place, again?" I prodded.

Evasively, he answered, "Trust me—you don't want to know. Anyway,

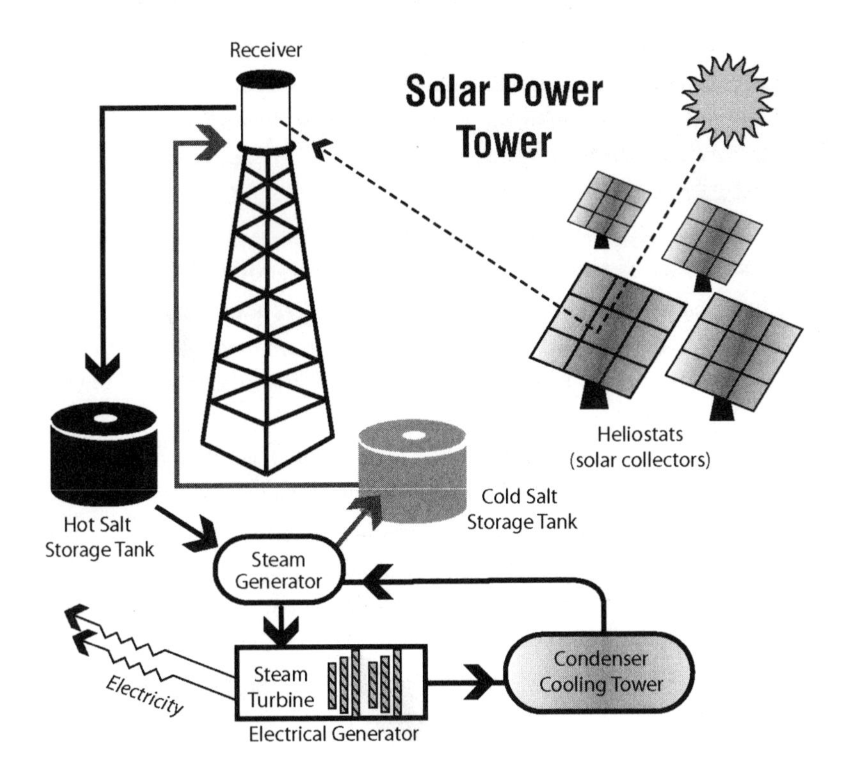

it's not important. Just don't get caught without a visa—at least not alive."

"*What??*"

"Just kidding, lad. Lighten up, will you?"

I wasn't so sure he was kidding, but I let it pass. "I take it this technology was developed first and foremost as a clean, renewable source of power for the electrical grid?"

"Right. But the nature of the design makes it eminently adaptable to hydrogen production. Unlike wind turbines, which can have long down times, a solar power tower is engineered to run up to 70 percent of the time on sunlight alone. The other 30 percent can be made up using relatively clean natural gas as a heat source. With the facility in fulltime operation, it would be a simple matter to devote a portion of its output to the production of hydrogen, through some form of electrolysis, or thermochemical water-splitting."

As Zed talked, I noticed the power tower and its components slowly returning to normal. First the fluids were again enclosed within pipes and tanks, then the bright blues and reds faded to black and silver. I took it to mean we were finished with this "attraction." I looked all around us; partly to see if there were any other larger-than-life demonstration facilities nearby, and partly to see if there were any armed insurgents coming over the hill.

"Is there anything else? Any other ways to make electricity from sunlight?"

Oddly, his expression became one of deliberation. As if he were considering just how much he wanted to tell me. What could be so secretive about the existence—or non-existence—of a different source of solar electricity? Finally, he said, "Well, yes. But it's back at the Wasserstoff Farm. In a sense, anyway."

Thoroughly confused and not wishing to extend our stay here if it wasn't necessary, I said, "Whatever. Can we leave?"

"Sure, why not? I'm grilling soy fajita burgers tonight."

Zed Pickett, soybean butcher. Oh, well...it almost sounded good.

technistoff - 14

For once, Zed's theme park attraction was smaller than the real thing. I'm sure it wasn't from any lack of ambition on Zed's part; maybe he simply didn't want to outshine the brilliant people at Sandia National Laboratories, Bechtel Corporation, The Boeing Company, Southern California Edison, and The National Renewable Energy Laboratory, who were the first to come up with the idea. Zed is a professional, above all else.

Solar Power Tower Two

The Saltern Tower of Power, as I later discovered, was based on a Power Tower facility, dubbed "Solar Two," an experimental power station that operated between 1994 and 1999 in the Mojave Desert near the town of Daggett, California. This remarkable power plant was a retrofit of Solar One, a solar power plant that operated from 1982 until 1988. The primary difference between the two power tower incarnations is that Solar Two used molten salt as a heat storage medium. Though Solar One produced over 38 million kilowatt hours of electricity during its six-year run, it used only water and steam to transfer heat from the receiver to the turbine, and so was severely limited in its capacity to store heat for nighttime production of electricity.

Solar Power Tower Two at Daggett, California. The tower with glowing receiver on top is surrounded by a field of heliostats. *Photo courtesy of Sandia National Laboratory.*

For all its success, Solar Two—with its 300-foot tower and 1,926 heliostats—was really a 10-megawatt prototype facility, built to demonstrate the viability of the technology. It has done this so successfully, in fact, that a third plant—aptly named Solar Tres—capable of 40-megawatts is currently under construction in Spain. Several other countries, including Mexico, Morocco, Egypt and India, are taking a close look at Solar Two's design.

power tower advantages

A power tower has two main advantages over an equally-sized photovoltaic solar array. Basing their conclusions on the cost and performance of a 200 megawatt facility, Sandia and NREL researchers concluded that a power tower can reap 17 percent of the annual harvest of sunlight for a cost of $0.06 per kilowatt hour, compared to $0.25 per kilowatt hour for a similarly-sized solar array using battery storage, and operating at a mere 10 percent annual efficiency. To learn more about power towers, read the Sandia National Laboratories 2000 "R & D 100" Awards Entry Form, at the Arizona Solar Center website: *www.azsolarcenter.com/links/faqs/solar.pdf.*

As Zed keenly observed, power towers—with their unequalled ability to store solar energy efficiently—are excellent candidates for hydrogen production. While all the power tower plants currently under consideration are being built for the purpose of supplying the grid with electricity to mitigate the greenhouse gases emitted by conventional fossil fuel-burning facilities, their use—either

A heliostat at Solar Power Tower Two shows the reflection of the tower. *Photo courtesy of Sandia National Laboratory.*

totally or partially—as hydrogen generating stations will not go unnoticed, once the hydrogen economy gets past the discussion stage.

solar power troughs

Though perhaps not as promising as the power tower concept, a couple of other heat-focusing technologies currently being explored deserve mention. Like the power tower, the power trough system heats a fluid (synthetic oil) which transfers its heat to water. The water, after being converted to steam, powers a steam turbine generator. No more complicated than a highly-reflective curved mirror with a black oil-filled tube mounted at the focal point, a power trough can reach annual sunlight-to-electricity efficiencies of 13 percent and should be able to provide electricity for around $0.12 per kWh (see the Arizona Solar Center reference link, above). The Achilles heel of this design lies in its inability to store heat for any appreciable length of time. This means the amount of supplemental fossil-fuel needed to run a power trough continuously is greater than for a similarly-rated power tower. Despite this drawback, nine different trough power plants in southern California currently supply 354 mega-watts of electricity to the power grid.

Concentrating Solar Power Troughs in Boron, California. *Photo courtesy of Sandia National Laboratory.*

Another interesting concept is the Stirling engine. Often called a dish/engine system, it uses an outside heat source to run a piston engine, very similar to a locomotive's steam engine. The engine, in turn, runs an electrical generator. The solar collector for the Stirling engine is at the focal point of a large, mirrored, parabolic dish. This system does have one advantage over the power tower and the power trough: since there is no steam turbine

Zed sez

Unlike wind turbines, which can have long down times, a solar power tower is engineered to run up to 70 percent of the time on sunlight alone.

used, a dish/engine system can be constructed from modular units, ranging in output from 5 to 50 kilowatts each.

solar power dish with Stirling engine

The lack of a steam turbine, it seems, would mean that a dish/engine system would be unable to use fossil fuels as a backup energy source, making it an unlikely candidate for hydrogen production. But all a Stirling engine really needs is heat, and this can be supplied by natural gas as easily as sunlight. It's just not as environmentally benign, or as philosophically gratifying.

For information about power troughs and dish/engine systems, plus some great power-tower graphics, visit the Boeing Company's website at: *www.boeing.com/assocproducts/energy/powertower.html.*

Concentrating Solar Power Dish with Stirling engine in Almeria, Spain. *Photo courtesy of Sandia National Laboratory.*

book three

★ ★ ★

PUTTING HYDROGEN TO WORK

Liquid Gas
Pipeline Rail Ship
Road Composite Tanks
H2 Glass Microspheres
Chemical & Metal Hydride
Hybrid Engines Fuel Cells
CH3OH Carbon Nanotubes
Vehicles Stationary FC
Internal Combustion
Home Heating
Portable

chapter 15

Stuffing Wasserstoff: Is There an Ideal Way to Store a Non-Ideal Gas?

The next day, as I looked from my cottage toward the small courtyard in front of the holosium, I found Zed in an unusual predicament. And while I was certain it was an entanglement he had somehow—willingly or otherwise—put himself into, it was no less odd. But he did seem to need my help, so rather than stand on the stoop and wonder how he had managed to get to where he was, I stepped into the day—warm and hazy, and breezy, as always—determined to set Zed free from a very peculiar prison.

I had no idea how to free a man who was trapped within an 8-foot glass sphere, but I felt certain that Zed must. He had, after all, gotten himself in there in the first place, unless I was to believe the Wasserstoff Farm had been invaded by giant, hollow, man-eating creatures of glass.

As I drew closer, my eyes began to make out more of the particulars that had escaped my attention from the distance of my cottage. For one thing, the sphere rested on a frame, set above a burner of some sort, though there was no flame presently burning to heat the sphere. And, as I walked up to the sphere, I saw that Zed was not alone; there were millions upon millions of near-microscopic particles—tiny blue doublets that could only be Zed's version of H_2 molecules.

I tried hard to stifle a chuckle when I saw the T-shirt *du jour*, which read: *You Can Order 'Hydrogen Under Glass' at The Wasserstoff Farm*. "It's a fine mess you've gotten yourself into this time," I told him.

Even through the curved surface of the glass, it was obvious by his anxious expression he was in no mood to listen to my assessment of his quandary. "Why don't you quit pontificating and light the burner!" he demanded, in a muffled voice. "I'm about to run out of air in here!"

While I couldn't see how his situation would be helped in the least by making things hotter, I lit the burner, just the same. The glass crackled as it began to warm and, after a few minutes, there began an exodus of H_2 molecules. Slowly at first, then faster as the glass warmed and they became more excited. When nothing but a few million stragglers remained, Zed made his move. First, he touched the side tenuously, then, seeing his fingers pass magically through the glass, he stepped out, as though the solid glass offered no more resistance than water.

Checking himself over to make certain he had made it out intact, he said, "Contrary thing sucked me right in."

Amused, I asked, "Care to explain?"

"Sure. You see, I was loading this oversized sphere with these, uh... 'special' hydrogen molecules to demonstrate a new storage technology, but when I turned off the burner, *whoosh!*—there I was, trapped like a rat."

"Sounds dreadful," I sympathized, "but what was it you were going to show me?"

"Well, you should know by now that one of the biggest obstacles to using hydrogen to power cars is in finding ways to store hydrogen in a small enough space. Otherwise, either the car won't go very far, or the occupants will have nowhere to sit."

"Could be a problem," I agreed.

He reached down and turned off the burner, then quickly scampered out of range. Satisfied the situation had been stabilized, he said, "Not 'could be', it is. And I'm sure you know why."

I thought for a moment, then answered, "Well, besides being the lightest element in the universe, hydrogen molecules have virtually no attraction for one another. That makes hydrogen very hard to condense, and even harder to liquefy."

His smile indicated that my answer met with his approval. He said, "Good, lad. You may get your book written, yet. You see, the irony with hydrogen is that it has more energy per pound than gasoline, or natural gas, or just about anything else you could name. But unfortunately, that pound takes up more space than anything else."

"How much more?" I asked.

"A lot. As a comparison, consider this: a pound of gasoline at normal temperature and pressure occupies about this much space..." With a twist of his wrist he produced a black cube a little over 3 inches on a side. It remained suspended in midair after he pulled his hand away. "Whereas a pound of hydrogen at normal temperature and pressure needs a little bigger container..." This time he turned his back to me and threw out his arms. Instantly another cube appeared from out of the ether, this one over 5½ feet high, nearly as tall as me.

I had to object. "Wait a minute, Zed. Don't we need to compare apples to apples, or in this case, Btus to Btus?"

"Quite right. Good point. And since a pound of hydrogen, with 60,000 Btu, has nearly 3 times the heat energy as a pound of gasoline at 20,800 Btu, the cube should really be this big..." With a wave of his arms, the hydrogen cube shrank down to just a little less than a 4-foot cube. He snatched the smaller gasoline cube out of the air and placed it on top of the big one. "Two 20,800-Btu cubes," he proudly proclaimed.

I looked at the two cubes with dismay. "What about compression?" I

Zed sez

The problem with hydrogen as either a gas or a liquid is that it requires heavy, expensive tanks to keep it cold, or pressurized, or both. But if you can find a way to bind hydrogen loosely to some other substance, you are no longer concerned about high pressures and frigid temperatures.

asked. "Can't you compress the hydrogen in the big cube to the volume of the small cube? I mean, isn't there some sort of...law...about that?"

"You mean the 'ideal gas law," he said. "Simply put, it states that whenever you double the pressure of an ideal gas, you halve the volume."

"Yeah, that's it."

"So, *if* hydrogen were an ideal gas, all we would have to do to make the cubes the same size is to put the hydrogen under enough pressure."

"So it would seem," I told him.

Zed replied, "But we have an obstacle, I'm afraid. First, *if* hydrogen were indeed an ideal gas, we would still have to put it under more than 38,000 pounds per square inch of pressure to make the big cube the size of the small one. A rather unwieldy amount wouldn't you say? But in reality, 10,000 psi is about the best you can hope for when compressing hydrogen. That's because it begins to deviate considerably from the ideal gas model beyond that pressure. And even at 10,000 psi—the pressure you'd feel pushing against you if you were covered with a little better than 4 miles of seawater—the hydrogen is only compressed to 65 percent of the density of an ideal gas. So, at 10,000 psi, our 4-foot cube shrinks to this..." He waved his hands and the big cube shrank down to a cube around 6½ inches on side.

"That's not so bad," I said.

"Not until you start stacking big cubes and little cubes into the same space. Then you'll see the difference. The little cube comprises around 38 cubic inches, the big one 284—over seven times as much space."

"So a car with a 15-gallon gas tank would need the equivalent of a...let's see...of a 105-gallon hydrogen tank."

"Bigger than two 55-gallon drums," Zed chimed in, "when you consider the thickness of a high-pressure tank."

"Houston...we have a problem," I lamented.

"And this doesn't even include the space taken up by pipes and pumps, and so on, to make it all work. Plus, the tank needs to be round for strength and even pressure distribution. It turns an automobile into a tank on wheels."

I thought for a moment, then said, "What about liquid hydrogen? Doesn't that take up less space?"

"Sure. Roughly half the space of high-pressure hydrogen—" he touched the bigger of the two cubes with his index finger, and it shrank again; this time to a cube a little over 5 inches on a side "—but it takes a cryogenic tank to keep hydrogen chilled to minus 253 degrees Celsius. These tanks are bulky and exceedingly expensive, and your hydrogen will boil-off at an unacceptable rate. Besides, the amount of energy needed to liquefy hydrogen is close to 30 percent of the energy of the hydrogen itself."

He rested his chin in the crook between his index finger and thumb, and said, "All in all, liquid hydrogen may turn out to be more trouble than it's worth."

I nodded. "The alcohol sisters are looking prettier by the minute," I told him.

"I never thought they were half-bad, myself," Zed agreed. He held his hand out in front of him, palms up, then made two fists. When he opened his hands again, they each contained a cube. The one in his left hand was just a little bigger than the gasoline cube—just shy of 4 inches—while the other was close to 4¼ inches. "Meet Ethanol and Methanol," he said, nodding first to his left, then to his right, "two room-temperature hydrogen carriers."

He set the cubes down next to the others, and said "But there's no need to throw in with them just yet. There are other ways to store hydrogen."

"Such as?"

"Well, lots of ways. The problem with hydrogen as either a gas or a liquid is that it requires heavy, expensive tanks to keep it cold, or pressurized, or both. But if you can find a way to bind hydrogen loosely to some other substance, you are no longer concerned about high pressures and frigid temperatures."

"Are we talking about solid hydrides?" I asked.

"We can if you'd like," he answered, "but they come with their own set of problems."

"What doesn't?" I wondered.

"Let's look at the bright side first, shall we? Since a solid hydride is simply a substance—usually a metal—with which hydrogen forms weak molecular bonds, the problems associated with pressure and temperature go away, for the most part. And so does our volume problem, since many types of hydrides store hydrogen more densely than liquid hydrogen."

"Sounds good," I said.

"Yeah, so far it does. But then we run into the 'too' problem. They're too heavy, or too expensive, or it takes too much heat to free up the hydrogen."

"Give me an example."

"Okay. Take lanthanum nickel hydride, for instance. Molecularly it's $LaNi_5H_6$, so you can see that it holds a fair bit of hydrogen. Better yet, add just a little pressure and it will release the hydrogen at nearly room temperature."

"So what's the problem?" I asked.

"Weight. Lanthanum nickel hydride is so heavy that it takes 1,000 pounds of the stuff to store 13 pounds of hydrogen. To store the equivalent energy of a 15-gallon gas tank, you would need nearly 2,400 pounds of it. Weight-wise, it makes about as much sense as carrying around your favorite marble inside a bowling ball."

This was getting frustrating. I asked, "Are there any other metal hydrides?"

Zed answered, "Sure. Dozens of them, all with their own problems. Right now, the most promising seems to be sodium alanate. Chemically, it's $NaAlH_4$. As you can see, with only one atom each of sodium and aluminum to every four atoms of hydrogen, it's going to be a lot lighter than lanthanum nickel hydride. In fact, theoretically it can hold 5.7 percent of its weight in hydrogen. That would make it only one-fourth the weight of $LaNi_5H_6$, so our 2,400 pounds of material would drop to a heavy—but much more manageable—600 pounds. In terms of volume, sodium alanate ranks about even with liquid hydrogen. Unwieldy, but workable, especially since you have more flexibility in terms of the container's shape."

"But...?"

"The main problem is that it takes a lot of time to recharge the material—rehydride it, if you will—even at high pressure. And, like all metal hydrides, it suffers the problem of contamination. Any impurities in the hydrogen go into the hydride, but they don't come back out. After awhile the storage capacity becomes noticeably diminished."

"C'mon, Zed, this is getting depressing. I need some good news here!"

"Don't we all? The problem is this: safe, efficient hydrogen storage was never that much of a problem until people started talking about increasing its use by several orders of magnitude and carrying it around in the family car where big pressurized tanks aren't practical. Believe me, there are armies of researchers working on the problem, night and day."

I released an audible sigh, and asked, "Well, what else have they come up with?"

"I don't suppose you've ever heard of sodium borohydride, have you?"

"No, not right off-hand," I confessed, "what is it?"

"It's a common material, rather like soap. Its chemical formula is $NaBH_4$, which is similar to the formula for sodium alanate, except that the relatively heavy aluminum atom is replaced by boron, which is even lighter than carbon. Best of all, it has about the same energy density, volume-wise, as gasoline, without being explosive or even flammable. And it can be stored in plastic containers at ambient temperatures and pressures."

"Sounds too good to be true," I said, waiting for the other shoe to drop, "but what do you have to do to get the hydrogen out of it?"

"Mix it with water," he answered.

"That's all?"

"In the presence of a chemical catalyst, of course. The catalyst is the key to controlling the reaction. In a car, the gas pedal would control a pump that would release $NaBH_4$ into the catalytic chamber, where hydrogen would instantly be produced and sent to the fuel cell. The $NaBH_4$, which has now become $NaBO_2$, or sodium perborate—very similar to household borax—is then pumped into a holding tank. When it comes time to refuel your car, $NaBH_4$ is pumped in one side, while $NaBO_2$ is pumped out the other. The $NaBO_2$ can then be rehydrided into $NaBH_4$, and delivered back to your car. Nothing is wasted."

"Terrific. What's the catch?"

"Simply that you are adding one more link in the chain. Rather than pumping pure hydrogen into a tank filled with an immobile hydride, you are actually removing the depleted hydride, recharging it—at no small expense—and pumping it back in. It's a somewhat inefficient way to dispense fuel. In a larger sense, you could say it encumbers the infrastructure. For that reason, many think it's an inelegant solution."

"To the devil with 'inelegant'. If it works, do it, I say."

"I think the same thing was said of 8-track tapes," Zed pointed out.

"Who cares?" I fumed. "Do it until something better comes along!"

Calmly, Zed said, "Retrofitting all of the world's cars and refueling stations after 'something better' comes along is a bit more difficult than discarding a worthless tape player."

He was right, and I knew it. It just seemed like such a pity to have to wait, while all the pieces of the hydrogen economy were so tantalizingly close to coming together. I said, "What else could possibly be out there? What's everyone waiting for?"

"The same thing everyone always waits for: a lightning bolt—the next great new idea," he answered.

"And what might that be?" I wondered.

Zed answered with a word: "Carbon."

"Carbon?"

"Right. But not just any old form of carbon. We're talking about carbon nanotubes."

I shook my head, hoping some of the old information would settle into the bottom of my brain and make room for more at the top. "Alright, I'll bite. What's a nanotube?"

"Well, if you remember back a few days, you will recall that carbon likes to form four bonds. It's what makes life possible. But it also means that there are any number of ways for carbon to bond with itself. Graphite is one form of pure carbon, diamond another. And then there are nanotubes."

"So you said, but what's a nanotube?" I asked again.

He stared at me blankly for a moment, then stomped over to a small shed beside the holosium. After a few moments of banging and clanging, he returned with a short piece of chicken wire. "This is graphite," he said.

"If you say so."

He then rolled it into a cylinder, bent the ends of the wire together and held it in front of me. "And this is a nanotube," he proclaimed.

"And you call yourself a wizard."

"No point in showing-off if it isn't necessary," he replied.

True. But it never seemed to stop him before. I said, "Not quite to scale, I take it?"

"Actually, depending on the nanotube—and the hair, of course—a nanotube can be anywhere from one 10,000th to one 50,000th the width of human hair. Rather small, in any case."

"I should say. But what's it got to do with hydrogen?"

"Hard to determine, at this point, but it is a fact that hydrogen can be stored in nanotubes at reasonable pressures, and easily released with a drop in pressure or an increase in temperature. Fantastic claims have been made for just how much hydrogen can be stored in the tiny little things, but it may turn out that their storage capabilities are adequate for our purposes. If they can achieve an energy density greater than liquid hydrogen—and there certainly is every reason to believe that they eventually can—then nanotubes just may be the on-board storage medium of the future."

For the better part of an hour we had been standing next to Zed's giant glass sphere. As it seemed that we had explored every other avenue of hydrogen storage, I rapped on the hard, clear ball with my knuckles, and asked, "I take it there remains one other possibility?"

Zed, I noticed, had been standing a safe distance away. Now he ambled closer and eyed the sphere with suspicion. Finally, he turned toward me, and said, "I never should have made it so big. It's hard to trust something that big."

I looked at the seemingly-harmless glass ball. A few million of Zed's 'special' hydrogen molecules still zoomed and zinged around inside. "How's it supposed to work?" I asked, unable to mask my skepticism.

Zed took a tenuous step closer, then said, "It has nothing to do with chemical bonding, as in hydrides. We're back to pressure here, pure and simple."

"I'm listening," I assured him.

"Okay, it's like this. The idea of storing hydrogen under pressure on-board a vehicle is just too tempting to walk away from. You don't have to worry about contaminating a hydride container, and eventually having to replace it. Or the tremendous weight of a metal hydride. But, as we discussed earlier, pressure tanks are bulky, expensive, and very limited in terms of shape.

"But what if you could store hydrogen inside billions of little high-pressure containers, then release it with the simple addition of heat? You could wrap your storage tank around your vehicle in any way you chose. And the walls of your tank wouldn't have to be much thicker than the walls of your gasoline tank."

"Glass spheres?" I asked, "like tiny versions of this—" I slapped the wizard-eating ball.

"Microspheres, actually," Zed replied, "with the feel, composition, and texture of extremely fine sand. Except without the rough edges. And hollow."

I narrowed my eyes, and asked, "Is this something you read about in Jules Verne?"

"Hardly," he answered. A little indignantly, I thought. "It's a simple concept. Heat the microspheres with hydrogen in a pressurized containment vessel. Hydrogen molecules permeate the walls of the microspheres and occupy the hollow spaces. Then you let things cool down. The pressurized hydrogen is trapped inside the spheres, and all you have to do to release it is to heat them up, again."

"Is this for real?" I was still skeptical.

"Scout's honor," he said to assure me, though it seemed unlikely that Zed had ever been a scout of any sort.

"But it's kind of 'out there', right?"

"A ways, but not as far out as nanotubes. And by storing your hydrogen inside billions of little glass balls, it does solve a lot of problems. Holding hydrogen at 10,000 psi, microspheres would rival liquid hydrogen, both in terms of weight and volume, once the bulk and mass of the cryogenic container was taken into account, and would require much less effort to maintain."

"But?"

"Like $NaBH_4$, it still requires another link in the chain. Rather than just refueling, you would be taking on, and taking off, microspheres, which would then have to be 'recharged' with heat and pressure. It might get complicated."

I studied Zed for the briefest of moments. He seemed put out by this sticky, complex problem of storing the simplest of elements. It was like he'd dragged a bag of gold for a thousand miles, only to discover that the chest he'd been assigned to carry it home across the Seven Seas was one-third the size needed for the job.

"It's a cool idea, Zed," I told him. And it was.

He smiled broadly, giving me a rare glimpse of his sparkling white teeth. "This is just the script. If you think it's cool, wait five years—you can see the movie."

I didn't bother to ask him what he meant.

technistoff - 15

Zed was never in any danger of suffocating. As he bent over to pull the chicken wire out of the shed, I saw him quickly pick up a glass cutter that had fallen out of the breast pocket of his lab coat. Or whatever he calls the strange garment he wears like a Druid's business suit.

But then I have to ask, *can Zed really create an illusion around himself so real he can't get himself out of it?*

I'll never understand sorcerers. Never.

storing hydrogen

On the other hand, hydrogen storage technology, however cryptic and arcane in its own right, does at least lend itself to scientific scrutiny. So we'll start at the top.

A graph showing how high-pressure hydrogen deviates from the ideal gas law can be found in February/March issue of *The Industrial Physicist* in an article titled "Bottling the Hydrogen Genie." At 10,000 psi, an ideal gas will achieve a density of over 55 grams per liter, while hydrogen falls below 40 grams per liter at that pressure. This article, by Frederick E. Pinkerton and Brian G. Wicke, both Technical Fellows at the General Motors Research and Development Center, methodically details the promises and pitfalls of current hydrogen storage technologies, and is available on The American Institute of Physics website at: *www.aip.org/tip/INPHFA/vol-10/iss-1/p20.html*. But this is hardly the deathblow for high-pressure storage tanks. Just because Zed thinks they take up too much space—and are unforgiving in their shapes—it doesn't mean they won't end up being "the solution" to store hydrogen fuel, at least in the short term.

High pressure tanks are currently the best option for storing pure hydrogen. *Photo courtesy of Quantum Fuel Systems Technologies Worldwide, Inc. ww.qtww.com*

DOE goal: travel 300 miles before refueling

One of the Department of Energy's (DOE) goals for carmakers is to produce a hydrogen-powered car capable of going a minimum of 300 miles without fueling. This is the mileage range for a gasoline-powered car averaging 20 mpg with a 15-gallon tank. But if hydrogen fuel-cell cars are indeed able to achieve the gasoline equivalent of 70 mpg (and it appears this claim is reasonable), then the tank need only hold the equivalent of 4.5 gallons of gasoline, or just a little over 9 pounds of hydrogen. At 10,000 psi, this much hydrogen can be stored in a 30-gallon high-pressure tank. Suddenly high-pressure hydrogen gas doesn't look so bad.

who is making fuel cell cars

For a comprehensive list of prototype fuel-cell vehicles, including mileage ranges, miles per gallon, and fuel type and method of fuel storage, visit the Fuel Cells website at: *www.fuelcells.org/fct/carchart.pdf*. A quick glance will show that most carmakers currently prefer compressed gas over other means of storage.

Another of DOE's goals—to be reached by 2010—is to develop a means for storage in which the weight of hydrogen is at least 6 percent of the total weight

GM's HydroGen3

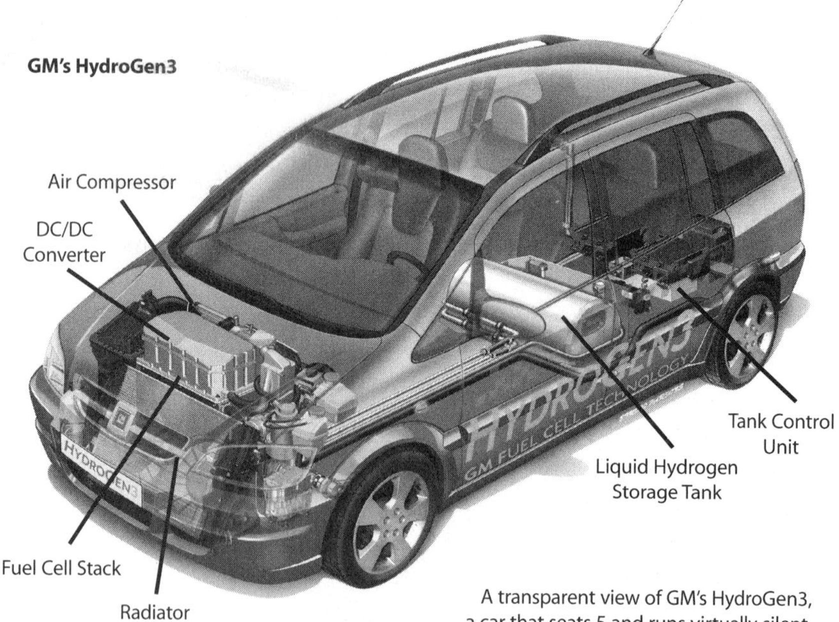

A transparent view of GM's HydroGen3, a car that seats 5 and runs virtually silent. Liquid hydrogen powers the fuel cell stack which supplies electricity to a battery, powering the vehicle and emitting only clean water. *Photo © General Motors. http://www.gm.com/company/gmability/adv_tech/400_fcv/index.html*

This hydrogen filling station, in the Opel test centre near Offenbach, Germany, was designed and built as a turnkey project by Linde Gas. *Photo courtesy of Linde Gas. www.linde.com*

(6 wt%) of the storage system. Lightweight high-pressure tanks, made of composite materials and lined with a bladder that reduces permeation of the H_2 into the walls of the tank, have already exceeded the DOE goal by showing a 12 wt% storage capacity at 10,000 psi. For details, see the DOE Energy Efficiency and Renewable Energy website section on hydrogen storage at: *www.eere.energy.gov/hydrogenandfuelcells/ hydrogen/storage.html*.

Nor is liquid hydrogen out of the picture. With the promise of more efficient means for liquefaction looming on the horizon, along with new techniques for storing liquid hydrogen at higher temperature and pressure, several automakers, including DaimlerChrysler, General Motors, Renault and Volkswagen, have developed liquid hydrogen-fueled prototypes (see *www.fuelcells.org*).

hydrides in cars

What about hydrides? Do they show any promise? While much work remains to be done before the technology is perfected or any one hydride emerges as the clear winner, the Japanese carmakers Honda, Toyota and Mazda have all developed prototypes that use metal hydrides as storage media, while DaimlerChrysler and Peugeot are giving the chemical hydride, sodium borohydride, some serious testing.

Of all the metal hydrides, the alanates are currently receiving the most attention. In a 2003 DOE paper, titled "Hydride Development of Hydrogen Storage," author Karl J. Gross and others outline the possibilities of using titanium-doped sodium alanates to achieve the DOE's 2010 volumetric and gravimetric goals for hydrogen storage. The team concludes that "solid progress is being made on the development of light-weight complex hydrides for hydrogen storage." This paper can be downloaded at the Energy Efficiency and

Renewable Energy website at: *www.eere.energy.gov/hydrogenandfuelcells/pdfs/iiib2_gross.pdf*.

A great deal of research is also being conducted on the chemical hydride, sodium borohydride. While the feasibility of using this abundant, hydrogen-rich

sodium borohydride

chemical as an onboard hydrogen delivery system has been successfully demonstrated (*www.eere.energy.gov/hydrogenandfuelcells/hydrogen/storage.html*) using a Millennium Cell-designed system installed in a Chrysler *Natrium* minivan (you can read about the *Natrium* at the H2 Cars Biz website, *www.h2cars.biz/artman/publish/article_105.shtml*), the cost of regenerating NaBH4 from $NaBO_2$ is proving to be expensive. In a short paper with a lengthy title ("Process for the Regeneration of Sodium Borate to Sodium Borohydride for Use as a Hydrogen Storage Source", available for download at: *www.eere.energy.gov/hydrogenandfuelcells/pdfs/iiid1_wu.pdf*), Dr. Ying Wu, Program Director, Synthesis, of Millennium Cell, Inc. discusses the possibilities of using hydrogen-assisted electrolysis to reduce sodium borate to sodium borohydride in a single-step process. We all wish him the greatest success.

carbon nanotubes

That brings us to the two storage technologies that have yet to be employed by any carmaker, namely carbon nanotubes and glass microspheres. Research on the ability of carbon nanotubes to adsorb and release hydrogen has been beset with difficulties from the beginning. Sample contamination and inconsistent testing procedures have yielded mixed, and even contradictory results. While researchers remain hopeful, it appears that, for the short term at least, the use of carbon nanotubes for practical hydrogen storage will have to take a backseat to more proven technologies. An interesting general discussion on the problems and promises of nanotube hydrogen storage can be found in the Pinkerton and Wicke article (cited above). For a more rigorous treatment of the subject, read Ragaiy Zidan's paper, "Doped Carbon Nanotubes for Hydrogen Storage," at the Energy

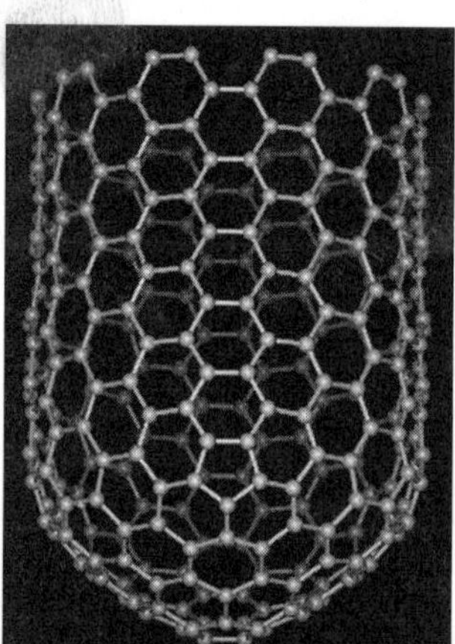

Nanotube representation. *Courtesy of The Smalley Group, Rice University.*

Efficiency and Renewable Energy website: *www.eere.energy.gov/hydrogenand fuelcells/pdfs/30535an.pdf.*

glass microspheres

And finally, the diminutive siblings of the silicon-based, non-life-form that swallowed my mentor: will glass microspheres ever prove to be a viable storage medium for hydrogen? While not nearly as complex as the problems facing carbon nanotube researchers, microsphere technology does, nonetheless, have its own unique difficulties.

The biggest problem is in finding just the right type of glass. Although it has been demonstrated that microspheres can be filled with up to 14,000 psi of hydrogen, they cannot hold that much pressure for long at ambient temperature. Generally, the purer the glass, the higher the permeation rates—both in and out. This is both good and bad, for while it is desirable to be able to retrieve your hydrogen quickly, your microspheres won't do you much good if all the gas leaks out before you want to use it. A compromise is required, wherein a type of glass is found that will allow adequate rates of hydrogen extraction with a minimum of leakage. By mixing the glass with impurities (such as sodium ions), the permeation rates can be adjusted, and researchers are hopeful of finding the magic formula that will take glass microspheres out of the laboratory and into the tank of your brand-spanking new fuel-cell-powered car. I'm sure Zed will be the first in line to try it out.

To learn a lot more about microspheres, read the 2001 paper, "On the Potential and Limitation of Hydrogen Storage in Microspheres." Written (in English) by Marius Herr, of the ET-Energie Technologie in Germany, and others, it can be downloaded at: *www.waterstof.org/20030805EHECO2-205.pdf.*

BULK STORAGE AND TRANSPORTATION

I should add a note here concerning bulk storage and transportation issues relating to hydrogen, since you may have noticed that Zed altogether avoided the subject. It wasn't just because he didn't think it was sexy enough; he simply believes these issues are far less acute than the onboard storage problems he did address.

Why? Because, simply put, it's easier to store things in large volumes. The volume of a sphere (or just about anything else) increases faster than its

Different ways of gas distribution—in gaseous form in cylinders (left) or large-volume delivery of liquified gas in tanker trucks. *Photo courtesy of Linde Gas. www.linde.com*

surface area as it grows in size. That means your balloon, or your elephant, will become a more efficient use of space as it grows. Large tanks to store hydrogen under pressure can store a much greater amount of hydrogen per square foot of surface area, making them eminently practical for bulk storage, both in terms of cost and ease of use.

Likewise, hydrogen can be moved as a low-pressure gas through existing natural gas pipelines, or transported in high-pressure tanks, either by truck, ship or rail, without the introduction of any ground-breaking new technology.

This is not to say that every problem related to bulk transport and storage has been resolved. Hydrogen is inherently costly and unwieldy to handle, and once the hydrogen economy picks up steam there will be many problems that will need to be addressed. But, hopefully, the technological advances made in the area of small-scale storage can be applied on the bulk level.

IS HYDROGEN SAFE?

Every time the subject of hydrogen comes up, someone asks the inevitable question, "Is hydrogen safe?" And usually I reply, "What are you, nuts? The stuff is rocket fuel, for crying out loud!"

It's my subtle way of letting them know they've asked the wrong question. Nothing is truly safe, and it strikes me as curious to hear this question asked by people who daily encounter gasoline, diesel fuel, LP gas, denatured alcohol, kerosene, paint thinner, acetone, or any of a dozen other common volatile concoctions that react violently with heat and flames.

A more informed question would be: *how safe is hydrogen compared to gasoline?* Or better yet: *how do the safety concerns of hydrogen differ from those for gasoline?* Now we're getting somewhere. After all, hydrogen's primary role will be as a replacement for gasoline, so it's the logical substance to compare it to.

And to be honest, hydrogen does have a few quirks that make it less safe than gasoline—in some cases. Take flame characteristics, for instance. Unlike a gasoline flame, a pure hydrogen flame is all but invisible. You might put your hand in a hydrogen flame, or even walk into it, before you knew there was a fire. This would not be a healthy thing to do. It has been suggested by some that a chemical could be added to hydrogen to give it a visible flame, but such a practice might not be practical if the hydrogen is being run through fuel cells that demand H_2 of very high purity.

The hydrogen detractors who like to point out the dangers of hydrogen's invisible flames, however, usually fail to acknowledge that, in most cases, the burning H_2 will ignite something around it that *will* give off a visible flame. This would most likely be the case in a car fire, for instance.

Even when it's not burning, hydrogen is undetectable to the senses, since H_2 gas has neither color nor odor. If your hydrogen-powered car developed a leak in its fuel system, you could walk into a garage with dangerous levels of hydrogen gas and not even know it. Until, of course, the pilot light on the furnace or hot water heater—or the match you use to light your smuggled Cuban cigar—ignites it. For that reason, hydrogen-powered vehicles will most likely usher in new standards for safety, including mandatory H_2 detectors—hardwired to vent fans that automatically kick in when unsafe levels of H_2 are detected—in any enclosed space where such vehicles might be parked.

Ford Focus FCV Hybrid hydrogen storage tank. *Photo courtesy of Ford Motor Company.*

Hydrogen is also considered dangerous by some because of the extreme pressures and/or cryogenic conditions under which it must be stored and transported. I tend to take the opposite view: the fact that such extravagant measures are employed to contain H_2 means that it will take near-cataclysmic circumstances to release it. The standards of safety set by the Department of Transportation for onboard hydrogen storage, for instance, are so stringent that your H_2 fuel tank may well be the toughest and most enduring component on your vehicle. Certainly if you were ever in a crash of sufficiently destructive energy to rupture your fuel tank, a burst tank would be the least of your worries.

What about the other side of the coin? What properties does hydrogen gas possess that make it safer than gasoline?

nontoxic

For one thing, hydrogen is nontoxic. There will never be an instance where hydrogen from a leaky underground storage tank poisons the ground—and the groundwater—around it for decades, as recently happened with a buried gasoline tank in the small rural town where I used to live. So severe were the health risks from more than 30,000 gallons of gasoline lost over a period of several years that the bank across the street from the filling station—the only bank in town—closed its doors…forever.

In fact, hydrogen, being the ethereal substance it is, is lighter than anything else in the universe. It goes nowhere but up when it finds a way out of the vessel that's holding it. So, while burning gasoline might well flow and slither across the ground like something out of a Dantean nightmare, burning H_2 forms a tall, lean flame that seems to want nothing more than a quick path to the sky. We should also remember that, unlike gasoline which creates a host of noxious chemicals when it burns, hydrogen creates only water.

And last but certainly not least, I should mention that it takes a fourfold higher concentration of hydrogen in the air to ignite, compared to gasoline. So, while hydrogen will sustain combustion when it comprises 4 percent by volume of the surrounding air, it takes only a one percent concentration for gasoline.

So, again: is hydrogen safe? It's a silly question; nothing that burns is safe. Is it safer than gasoline? Only time will answer that question definitively, since we humans are so adept at finding creative new ways to hurt ourselves, but I'd wager that the safe money is on hydrogen.

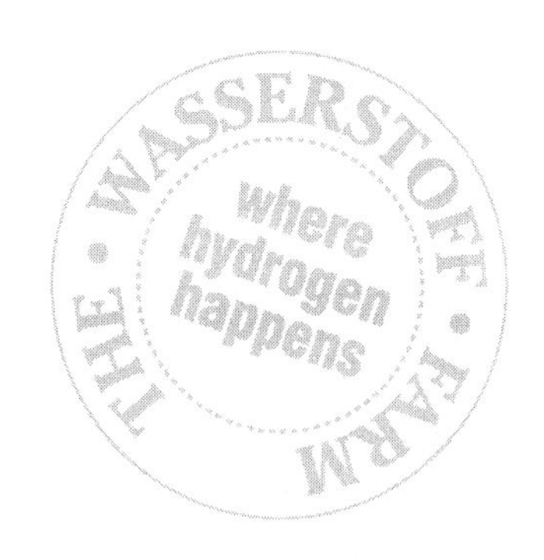

chapter 16

The *Infernal* Combustion Engine: Can the Ol' Hoss Run on New-fangled Feed?

As I eyed Zed's wondrous glass sphere and pondered all the different hydrogen storage options he'd run by me, the old magician disappeared. Literally. It might simply have been a case of him sneaking off while I was lost in a daydream, or it may have been another of his illusions. But either way, he was gone.

So there I was, alone in the middle of a day without a whole lot to do. My eyelids start to droop and I felt my energy level dip near the hibernation range. Rather than worry about where Zed may or may not have gone, I ambled back toward the cottage for a little nap. But no sooner had my hand grasped the doorknob, than all hopes of a midday respite scurried into the great cosmic void.

The deep, throaty rumble of a V-8 engine is like a lion's roar, or a coyote's howl, or a snake's rattle—it's primal. At least for anyone

who, like me, grew up around guys who made their horses stand out in the rain so they could work on their muscle cars in the comfort of a barn. And to hear one of these engines now—the last sound I would have expected to hear at the Wasserstoff Farm—was enough to dissolve any thoughts of sleep that had begun to crystallize in my now-attentive brain.

I turned quickly to see Zed—wearing wrap-around goggles—pull around the far side of the greenhouse in the coolest little metallic-blue roadster I'd ever seen. But what was it? It kind of resembled a '40s vintage coupe of some sort, but the classic lines had been modified; as if an old idea had been rethought with a space-age twist. Then I noticed the grill, where "PMC" was artfully displayed in gold lettering, and I knew.

Pickett Motor Company, indeed.

Zed pulled right up to my door, and hollered, "Hop in! Let's go for spin!"

I didn't need to be asked twice. The amply stuffed, tuck 'n button Naugahyde bucket seat, I quickly noticed, conformed perfectly to my body. This was good, because the second I got settled in Zed put the pedal to the metal and sent me so far into the seat I thought my backbone would fuse with the springs.

Like a high school kid with more horsepower than common sense, Zed screamed around the holosium, ripping up grass and dandelions, sending rocks and gravel flying in every direction. It brought back bittersweet memories of my ill-spent youth.

Once beyond the buildings, he found a straightaway between the trees and bushes and opened it up. Supposing the car was running on hydrogen, rather than gasoline—the rumble seat, I'd noticed had been replaced with a shiny, thick-walled tank—I listened for backfiring, or general roughness in the sound of the motor, since I had heard that these were standard problems with internal combustion engines converted to run on hydrogen. But this engine suffered no such problems; it never missed a lick, or offered up a single errant sound. Motors at the drag strip should be so well tuned.

Zed accelerated quickly through the gears, then held it steady as he bolted over the formerly-tranquil plain of the Wasserstoff Farm. The engine purred like a 1,000-pound kitten.

Satisfied he'd made his point, he turned the car in a wide circle and took it back to the holosium, parking in the graveled area where the giant glass sphere had been, but mysteriously no longer was. I quickly jumped out to examine the engine, openly visible since the old-fashioned hinged hood flaps had been removed to make it look like a street rod.

Zed stepped out the other side and watched me, amused, as I tried to make sense of the shiny powerhouse with water dripping from the eight unmuffled exhaust pipes. "Looking for something?" he asked.

I looked up and noticed for the first time that he'd changed his T-shirt. The new one, featuring a picture of his hot rod, read: *Hydrogen Roars at The Wasserstoff Farm*. Interesting, but not nearly as interesting as his car. I asked, "How'd you do it? How did you make this engine run so good on hydrogen?"

Quizzically, he said, "Why? Do you suspect magic?"

"Well, you know, that's always a real possibility around here."

Zed grinned conspiratorially. "Always and forever. But to be honest, this is nothing but good engineering.

"You see," he went on, "contrary to what you may have heard, pure hydrogen is an excellent fuel for internal combustion engines. It can be combusted over a much wider range of fuel/air mixtures than gasoline, meaning that it can be used in a 'lean' mixture. This makes for more complete combustion and greater fuel economy.

"Besides that, hydrogen has a very low ignition energy—around ten times lower than gasoline. It allows for easy starting, even when running on lean mixtures."

"Yeah," I piped in, "but doesn't that have drawbacks? I mean, if hydrogen ignites so easily, don't you run into the problem of 'pre-ignition', when the cylinder gasses ignite before they're supposed to?"

"It's fixable."

"And what about backfire, when the pre-ignited flames in the cylinder

Zed sez

Contrary to what you may have heard, pure hydrogen is an excellent fuel for internal combustion engines. It can be combusted over a much wider range of fuel/air mixtures than gasoline, which makes for more complete combustion and greater fuel economy.

shoot through the intake valve before it closes, and ignite the gases in the intake manifold?"

"A minor problem," he smugly answered.

"I'm listening," I told him.

"Very well. But before I begin I want to be sure you know how a four-cycle engine works."

"I do," I assured him. "Me an' V-8s go way back."

"Prove it," he said.

"Okay, let's see here. Each piston is attached by a connecting rod to the crankshaft, which turns as the piston goes up and down. In a four-cycle engine the piston goes through four strokes—two up and two down—to complete a cycle. On the intake stroke the piston goes down as the intake valve opens and lets the fuel/air mixture into the cylinder. At the bottom of the stroke the valve closes and the piston rises, compressing the fuel/air mixture. When the piston reaches the top, the spark plug ignites the hot, compressed gases and forces the piston down in a power stroke. As the piston heads back up, the exhaust valve opens, allowing the spent gases to escape during the exhaust stroke. During each complete cycle the crankshaft makes two turns.

"There," I said, taking a breath. "Satisfied?"

"Very good!" he exclaimed. "Considering your knowledge of the subject, this should be easy. We'll address the backfire problem first. You see, there are several ways to deliver the air/fuel mixture into the cylinder. With a conventional carburetor or a central injection system, for instance, the combustion gases pass through the intake manifold from a central point, and into the cylinders by way of the intake valves. It's rather like feeding eight suckling calves by pouring milk through a hole in the top of a single container with eight nipples. If a calf burps while biting on an open nipple, its stomach gases pass back into the main container. This is no problem for calves, of course, but it certainly can be for the hapless motorist.

"An improvement on this theme is the port injection system, where the fuel is injected separately from the air, at each intake port. In this way there are fewer volatile gases in the manifold, since its main purpose now is just to supply air. This is something akin to having eight syringes filled with milk, attached to the eight nipples. When the calf bites down to open the nipple—analogous to the intake valve opening on the intake stroke, as the piston is going down—milk is sprayed into its mouth. Of course (leaving our calf to nurse on his own for the moment) as long as the intake valve is open with combustible gases in the cylinder, there is always the chance of backfire."

"Understood," I said, examining the injection system on Zed's roadster. It looked different from anything I'd ever seen before. "So what do you do about it? It's not like you can just close off the intake valve. It has to open, doesn't it?"

"To let in air, perhaps, but not necessarily to let in fuel."

"Huh?"

"Look at it this way. When the piston goes down, the intake valve opens, to let fuel and air into the cylinder, right?"

"Right."

"And when the piston comes back up on the compression stroke, all the valves are closed."

"Of course," I agreed.

"So, all we have to do to avoid backfire is to inject the fuel—separately from the air—into the cylinder when it's coming back up on the compression stroke. With no valves open, there is no possibility for backfire. It's called 'direct injection'. Pretty cool, idea, huh?"

"You can *do* that?"

"Sure. You just have to inject the fuel at a very high pressure. But it's both possible and practical."

It was a great idea, I had to admit, but didn't that only solve half the problem? "What about pre-ignition?" I asked. "Don't you still run the risk of the gases igniting inside the cylinder before the piston reaches the top of the compression stroke? That can really make for an inefficient, rough running engine."

He graced me with that mock-insulted expression of his, and asked, "Did it seem to you that *my* engine was running rough—or inefficiently?"

"No, but..."

"Alright, then."

"But how do you do it?"

"You can do it with water," he said, as if any idiot would know that.

"Come again?"

"Pre-ignition is generally caused by hot spots within the combustion

Zed sez

What I'm trying to tell you, lad, is that a properly designed hydrogen engine will have 20 percent more power than a comparable gasoline engine, and will be proportionately more efficient.

chamber, resulting from the incomplete mixing of gases prior to combustion. One of the best ways to cool down those hot spots is to mix water with the hydrogen, prior to injection into the cylinder. You can also recirculate part of your exhaust gases back through, but that cuts into your power."

Scratching my head, I said, "And that's all you have to do to make a gasoline engine run well on hydrogen?"

"Well, if you want *really* good performance, you won't try to retrofit a gasoline engine; you'll build a hydrogen engine from the ground up. That means redesigning the combustion chamber with a flat piston and cylinder head to reduce turbulence. You should also provide two spark plugs per cylinder for more even combustion, and two exhaust valves to clear the exhaust gases more quickly and efficiently. And if you can design a system that forces fresh air into the combustion chamber on the exhaust stroke, your chances of pre-ignition will be minimal—with no need of water."

"Is that all?"

"Not quite. To make your engine *optimally* efficient, you'll give it a large bore-to-stroke ratio and a much higher compression ratio than a gasoline engine, since, unlike gasoline or diesel, the heat of compression does not cause hydrogen to ignite prematurely."

Nothing he had said sounded in the least bit difficult, at least not from a manufacturing standpoint. It was just a matter of engineering. But I had always heard that internal combustion engines running on hydrogen were less efficient and less powerful than gasoline engines. I asked Zed, who answered, "Take a gazelle and a cheetah. They're both fast, and both efficient. But make them eat each other's food and see what happens. If you pour gasoline into an engine designed to run on hydrogen, you'd be very disappointed.

"What I'm trying to tell you, lad, is that a properly designed hydrogen engine will have 20 percent more power than a comparable gasoline engine, and will be proportionately more efficient."

I was impressed. I had no idea it was possible to get such performance out of a hydrogen-fueled internal combustion engine. Still, I was skeptical. "You know, that's really great, Zed, but 20 percent isn't so much, once you stop to consider that fuel-cell powered cars are expected to get the equivalent of 70 miles per gallon of gasoline."

He thought about that for a moment, then replied, "True enough. But there are two things you may have overlooked. First, that fuel cells are still too bulky and expensive to be practical for anyone but the well-heeled, and second, there is already a proven technology in mass production that can double the gas mileage of an internal combustion engine."

He was right. "You mean the so-called 'hybrid' cars, I take it?"

"Exactly. By combining the torque and power of a standard internal combustion engine with the efficiency and steady operation of an electric motor, along with clever ideas like shunting the force generated during braking to an electric generator, you can squeeze every last ounce of power out of your fuel. And if you use a hydrogen-powered engine in your hybrid instead of one powered by gasoline, you could come close to rivaling the efficiency of a fuel cell. And you could do it today."

"If you could just find a place to fill up," I added.

"Quite right. Shall we take another spin around the old Wasserstoff Farm?"

"I'd love to," I told him. "Can I drive, this time?"

"Just take it easy on an old man, okay?"

I rolled my eyes. "Sure thing, Pops."

technistoff - 16

I was aware that internal combustion engines could run on hydrogen long before I ever met Zed. What I didn't know was how *well* they could be made to run. I had never heard of a direct injection system (where the hydrogen gas is injected into the cylinder after the intake valves are closed), or using water to cool combustion chamber hot spots to solve the problem of pre-ignition. Certainly I had no idea that modifications such as changing the geometry of the combustion chamber, increasing the bore-to-stroke ratio, or adding extra valves and spark plugs could make such a radical difference.

hydrogen internal combustion engines

Still I had to be sure, since the art of illusion is part of Zed's nature. On the DOE's Energy Efficiency and Renewable Energy website, I luckily discovered a document simply titled, "Module 3: Hydrogen Use in Internal Combustion Engines." It was part of a course for Hydrogen Fuel Cell Engines and Related Technologies, conducted by College of the Desert, in Palm Desert, California. (This document can be downloaded at: *www.eere.energy.gov/hydrogenandfuelcells/fuelcells/pdfs/fcm03r0.pdf.*) Unless you plan on becoming an automotive engineer, you'll find everything you'd ever want to know about making an internal combustion engine run on hydrogen.

For a book that covers much of the same territory, as well as lots of hints for the dedicated backyard mechanic who wants to convert an old (or a new) car to run on hydrogen, see Michael Peavey's book, *Fuel from Water: Energy Independence With Hydrogen*. In addition to the technical information, this book

The Ford Focus C-MAX hydrogen internal combustion engine experimental vehicle. *Photo courtesy of Ford Motor Company.*

offers numerous examples of vehicles that have been converted to run on hydrogen.

Currently, three auto-makers (Ford, Mazda and BMW) are experimenting with hydrogen-fueled internal combustion engines. The Ford Focus H2ICE (*Hydrogen Internal Combustion Engine*) uses a four-cylinder, 16-valve, 2.3 liter engine, fueled by hydrogen gas compressed to 5,000 psi *(see photo)*. Performance-wise, it is said to match the power output of a comparably-sized gasoline engine, with up to 25 percent better fuel economy. You can read about the H2ICE at: *http://media.ford.com/newsroom/release_display.cfm?release =16375*.

Mazda and BMW are both testing dual fuel (hydrogen and gasoline) engines. Despite the use of direct injection and a pair of hydrogen injectors per each combustion chamber, Mazda's RENESIS is still only able to achieve 54 percent power when it's running on hydrogen, compared to gasoline (81 kilowatts with hydrogen, 154 kilowatts with gasoline). An October 2003 article about this car can be downloaded at the Auto Web website: *www.autoweb.com.au*.

dual fuel engines: hydrogen and gas

BMW 745h prototype with hydrogen internal combusion engine. © *Copyright BMW (GB) Ltd.*

BMW's 750hL liquid hydrogen-fueled sedan, by contrast, shows a great deal of promise, though in its current incarnation it's a prime example of what happens when you mix cheetah food and gazelle food together, and feed the heterogeneous concoction to both critters. Being tuned to run on both fuels, it isn't optimally tuned to run on either. The 12-cylinder 750hL produces 204 horsepower in dual-fuel form, but it's capable of 380 horsepower when tweaked to run on hydrogen alone; something BMW would very much like to do, once liquid hydrogen becomes universally available. An article by Ron Cogan about this car (*Running on Hydrogen, Driving Toward the Future*) can be downloaded at the Valvoline Car Care website: *www.valvoline.com/carcare/*.

ETHANOL & METHANOL: A NEW VENUE FOR THE ALCOHOL SISTERS?

What about the alcohol sisters? Will they ever get the chance to prove themselves as fuel for internal combustion engines? While many will raise the objection that the burning of both methanol and ethanol puts carbon dioxide back into the atmosphere, if these fuels are produced from biomass (a natural carbon sink) and the carbon dioxide released during their production is sequestered, both can be burned in internal combustion engines (or fuel cells, for that matter) with a *net decrease* in atmospheric CO_2 (see Chapter 11).

ethanol in Brazil

But are ethanol and methanol any good as fuels for internal combustion engines? Although both fuels have less energy density than gasoline (methanol about half that of gasoline; ethanol a little better at around 63 percent), with certain modifications either fuel can outperform gasoline in terms of power. Methanol has already proven itself as a fuel for race cars, while ethanol is increasingly becoming the fuel of choice for Brazilians who, with an abundance of sugar cane from which ethanol can be easily distilled, hope to be the first country in the world in which 100 percent of the cars run on 200 proof shine.

engine modifications

Perhaps the biggest problem of using alcohols as automotive fuels is their high heat of vaporization. Simply put, if your engine is much colder than room temperature when you turn the key, you probably won't get your car started with straight alcohol.

Many schemes have been devised to get around this problem—electric block heaters being about the most impractical of the lot—but the best solution is to start the car with a more easily vaporized fuel, such as propane, then switch back to alcohol, once the engine warms. In very cold climes, heat from the exhaust can be circulated over the intake manifold to keep the alcohol from condensing on the inside walls of the manifold.

And since it takes a greater volume of alcohol for an internal combustion engine to run properly, the fuel system has to be modified to deliver more fuel to the combustion chamber. Generally this problem can be fixed with a bigger fuel pump and different carburetor jets, or injector nozzles. Websites for learning more about methanol and ethanol as automotive fuels include: Transport Canada, at *www.tc.gc.ca/roadsafety/atvpgm/tec03.htm*; the Green Trust website at *www.green-trust.org*; and the American Lung Association of California website at *www.californialung.org/spotlight/cleanair03-alt.html*.

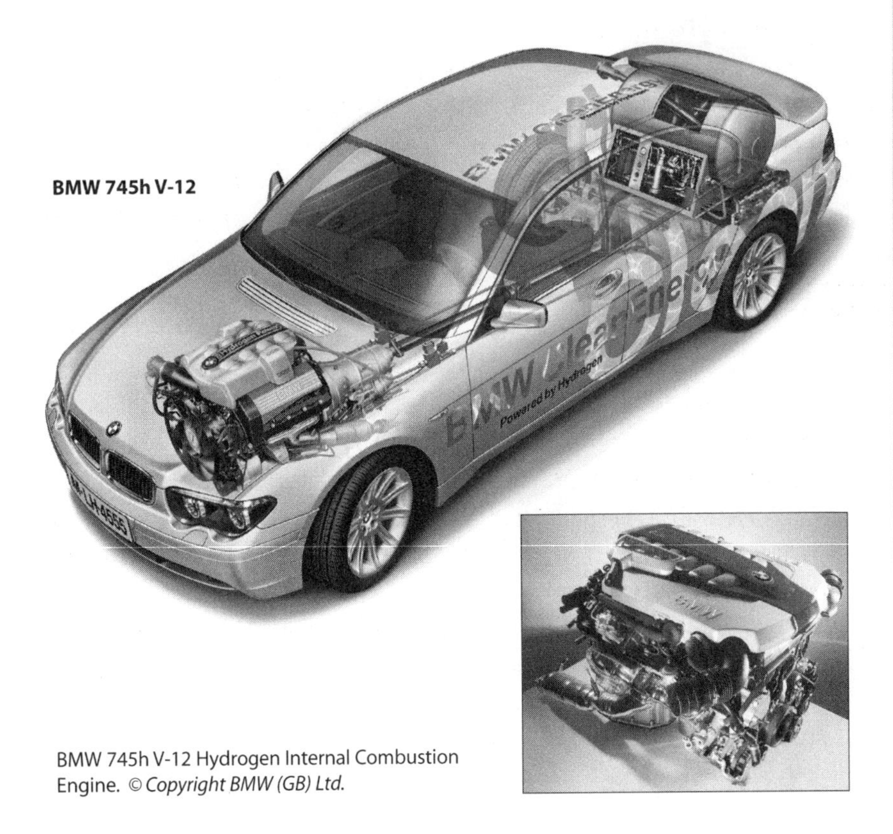

BMW 745h V-12

BMW 745h V-12 Hydrogen Internal Combustion Engine. © *Copyright BMW (GB) Ltd.*

A FINAL NOTE: one nagging problem associated with internal combustion engines is their annoying tendency to cleave atmospheric nitrogen (N_2) and transform it into nitrogen oxides (NO_x), common forms of automotive pollution. Because of hydrogen's low heat of ignition, a well-designed hydrogen engine will create fewer nitrogen oxide compounds and, with the proper catalysts incorporated into the exhaust system, it should be possible to remove the lion's share of what remains before it ever reaches the end of the tailpipe, or so says Ford Motor Company: *www.ford.com/en/innovation/engineFuelTechnology/hydrogenInternalCombustion.htm.*

reduced nitrogen oxide emissions

And you thought the ol' hoss' day was done.

FUEL ACRONYMS

H_2	Hydrogen
M100	Methanol
M85	85% Methanol, 15% Gasoline
E100	Ethanol
E85	85% Ethanol, 15% Gasoline
RFG	Reformulated Gasoline
LPG	Liquid Petroleum Gas (mostly propane)
CNG	Compressed Natural Gas

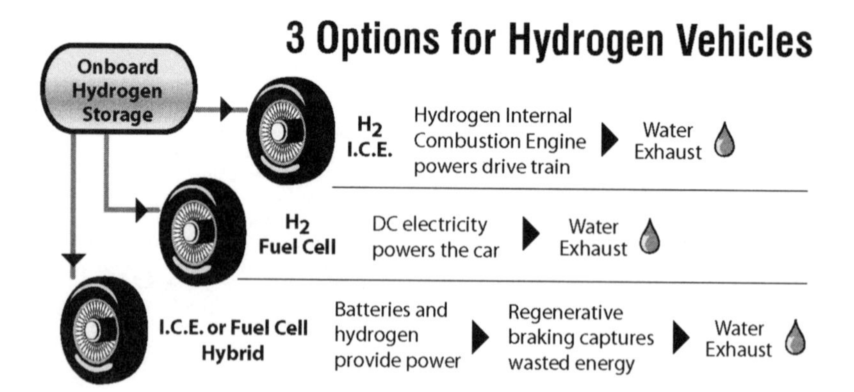

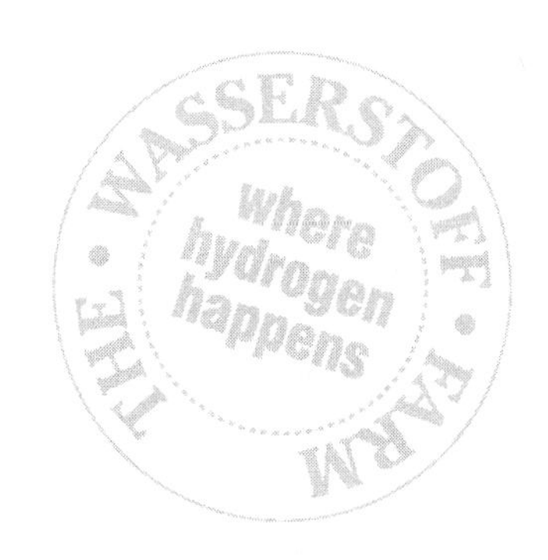

chapter 17

Fuel Cells: Going Where No Engine Has Gone Before

Zed shook me awake at the crack of dawn. "Get up! Get up! We've got a big day ahead of us!" This was bad news. I had tossed and turned most of the night, trying to shake off a tenacious nightmare, similar to the one I'd had my first night in the cottage. After finally coaxing Mike to the center of the Wasserstoff Farm, where I was just about to unravel its mysteries, we were abducted by glowing-eyed creatures from the planet *Fuelcellica.* They wanted to atomize me and my horse in a new, experimental "bio-cell." By the time Zed tried to roust me, we'd finally escaped in a cosmic-ray-fuel-cell-powered ship and found refuge with a sympathetic subterranean civilization living in a honeycomb of giant caverns beneath jungles of New Guinea.

I vowed then and there to spend less time watching the cartoon show, "Duck Dodgers, of the 24th and One Half Century."

Zed sez

Okay, so tell me—what exactly is a fuel cell?

Exactly, a fuel cell is a fuel cell. Inexactly, it's like a battery, or an engine, or even an electrical generator.

The truth is, a fuel cell is in a class by itself.

I was hardly ready for a big day, but Zed wasn't taking no for an answer. He pushed and prodded and harangued. And then he fought dirty—he waved a cup of his excellent coffee beneath my nose. That did it. I grabbed the cup out of his hands and drank half of it down without even noticing it was thermally unfit for human consumption.

I looked at his T-shirt—*Get Off Your Carnot—Fuel Cells Rule!*—and said, "Let me guess...today we're going to learn about fuel cells."

"No. *You* are going to learn about fuel cells. *I'm* going to teach you. I thought you'd have that straight by now."

"Whatever," I groaned, rolling out bed and slipping into my jeans, "I'm sure it will be yet another spellbinding safari into the deep, uncharted jungles of pedagogy."

I must've begun to annoy him. Sharply, he said, "Your lack of enthusiasm would be contagious, were there anyone here to catch it."

"Sorry, Zed. Bad night. I'm sure you put a lot of work into today's lesson."

"You don't know the half of it, lad. Now let's get moving."

Once outside, I looked carefully around before stepping off the stoop. Things at the Wasserstoff Farm had a way of changing in the middle of the night. Everything seemed normal...except for the stylish yellow sports car, sitting where the vintage coupe had been last night. Suddenly, the day was looking brighter. I jumped off the stoop and made a beeline for the car.

Zed grabbed me by the arm, and said, "Hold your horses! You shouldn't get in an all-fired hurry to drive something you know nothing about."

I stopped in my tracks, and said, "You mean fuel cells?"

"And everything that goes with them," he added.

Eager to dispense with the preliminaries, I said, "Okay, so tell me—what exactly *is* a fuel cell?"

"*Exactly*, a fuel cell is a fuel cell. *Inexactly*, it's like a battery, or an engine, or even an electrical generator. The truth is, a fuel cell is in a class by itself."

"I'm listening," I assured him.

"You see," he went on, "a fuel cell is like a battery in that it takes the energy of a chemical reaction and transforms it into electricity, but it's not quite like a battery, because it doesn't store any energy—it simply makes it.

"In that sense, it's like an engine, because it uses air and fuel to produce energy, and gives off heat and exhaust as byproducts. Yet it's different from an engine because it doesn't use fuel to create any mechanical energy. At least not directly.

"And, like an electrical generator, a fuel cell produces electrical power, though it uses fuel directly, rather than the power produced by fuel that's burned in an engine, or power from fuel-heated steam, or wind, or water."

He concluded by adding, "Frankly, lad, there really isn't anything quite like a fuel cell—except, as I said, a fuel cell."

It was a good answer, but I needed to know more. "But can you explain why I would *want* a fuel-cell vehicle, Zed, when you just showed me yesterday how efficient and non-polluting hydrogen-fueled internal combustion engines can be?"

He said, "For starters, things that don't have moving parts tend to be more reliable than things that do. But that's not the main reason. The fact is, the most efficient fuel cell may someday be twice as efficient at converting fuel into energy than the most efficient internal combustion engine."

"Why is that?" I wondered.

"Ever hear of Carnot efficiency?"

"Sort of," I said, eyeing his T-shirt.

"Simply stated, the Carnot efficiency of a heat engine—such as the big V-8 in that vintage roadster you nearly totaled last night..."

"Uh, sorry, Zed. I think the gas pedal got stuck."

"...is theoretically limited by the difference between the heat it converts into mechanical energy, and the heat it sloughs-off as waste. Unlike internal combustion engines, however, fuel cells don't convert chemical

Zed sez

Unlike internal combustion engines, however, fuel cells don't convert chemical energy into heat energy, and then into mechanical energy; they simply convert chemical energy into electrical energy. They can thumb their noses at Carnot.

energy into heat energy, and then into mechanical energy; they simply convert chemical energy into electrical energy. They can thumb their noses at Carnot. Even at this stage of the game—when fuel cells are far from being perfected—they're more efficient than internal combustion engines. And the gap can only get bigger."

"A power supply with a future?"

"Well said, lad. Let's take a look at the car."

"Should I pop the hood?"

"Oh, I think we can do better than that." He reached down under the grill, undid two latches, and lifted up the whole body. It was hinged in back and a pair of folding corner braces held it in place, like the legs of a card table. "It's not a production model," he confided.

"Nothing ever is, around here," I remarked, looking for, and failing to find, anything on the exposed chassis that looked familiar. "Where do we start?" I wondered.

"How about with the fuel tank? It seems logical." Zed rested his hand on a large, cylindrical tank with rounded ends. "It has dozens of layers of spun carbon fibers over aluminum. It holds hydrogen compressed to 10,000 pounds per square inch."

"Is this the fuel cell, underneath the fuel tank?" I asked, pointing to a large, rectangular box with high-pressure hoses running into either end.

"That's it. A PEM fuel-cell stack. It's the powerhouse that runs the car."

It was smaller than I thought it would be—maybe 36 inches by 18 inches, and a foot or so thick. I asked, "What's PEM stand for?"

"Proto-matter Evolving Matrix," he said, deadpan serious.

Just a little too serious, I thought. "I'm not buying it, Zed. Care to try again?"

He rolled his eyes and said, "Okay, try this: it's either 'Proton Exchange Membrane', or 'Polymer Electrolyte Membrane'. Take your pick—it ends up 'PEM' any way you slice it."

"For real?" I asked, still suspicious.

"Honest."

"Nothing like a dual-purpose acronym. It's nice and convenient."

"So it is," he agreed. "Instead of a liquid electrolyte like other types of fuel cells, a PEM fuel cell uses a thin plastic-like membrane, no thicker than a few sheets of paper. That's the 'polymer electrolyte' part. And it also allows positively charged protons to pass through, from one side to the other, once they're stripped of their electrons. Ergo, 'proton exchange'."

"But how does it work?"

"We'll get to that soon, lad. Right now we'll run through the basics. As you know, the fuel cell doesn't run the car directly—it just makes electricity. Direct current, or DC, electricity. However, the induction motor that turns the wheels," he pointed to a heavy case mounted over the front wheels, "runs on 3-phase AC."

"This is beginning to sound like our wind turbine/electrolyzer setup, except in reverse," I observed.

"Very much so," Zed agreed. "Our wind turbine uses mechanical energy to produce 3-phase AC, which is then converted to DC, the type of current needed to run the electrolyzer that turns water into hydrogen and oxygen. Here, with this car, we're using hydrogen and oxygen in a fuel cell—a reverse electrolyzer, if you will—to produce DC. We then convert the DC into 3-phase AC, and use it to produce mechanical energy. It's the flipside of the coin. If you think of the drive wheels as a pair of propellers, the reverse analogy is nearly perfect."

"Nearly?"

"Well, yes. The wind turbine produces nearly 700 volts AC, part of which has to be stepped down to less than two volts DC to run the electrolyzer. Here, we have a fuel cell stack that puts out over 250 volts DC, and an AC motor that runs on nearly the same voltage. The voltage conversion is less dramatic, you see."

"That's nice," I absently commented, entranced as I was by all the strange components of this futuristic sports car. "What's this?" I asked, pointing to a large, complex looking unit, sitting on top of the motor.

"That's the compressor. It sends compressed air to the fuel cell stack."

"Why? I mean, why does it have to be compressed?"

"Because it increases the efficiency of the cathode reaction, which is a real limiting factor. This is because the cathode reaction is several times less efficient than the anode reaction."

"Careful, Zed," I warned. "You just might teach me something about fuel cells."

"Oh, indeed I will, as you'll see...shortly."

I didn't like the way he said that. I couldn't help but think he had something planned that I wouldn't willingly agree to. I made a mental note to keep my guard up, especially if he tried to lure me anywhere near that surly dimensional compressor.

For now, I continued scanning the drive train and its auxiliary components. Specifically, I was looking for a transmission, which I was unable to find. I asked Zed about this. He replied, "It has a single-speed transaxle; one gear forward, one in reverse."

"But why not a transmission?" I asked.

"Doesn't need it. The electric motor can turn over 12,000 revolutions per minute. Most internal combustion engines are rarely taken above 4,000 rpm. So, with an average rpm range over three times that of an internal combustion engine—added to the fact that the horsepower of an electric motor is not as dependent on rpm as an internal combustion engine is—it's more efficient just to make the motor turn faster, than to add the extra inertia of a transmission."

A good idea, I had to admit, though I felt I would miss the fine art of finding the right gear, at just the right time. "Can we try it out?" I asked, eager to see how it maneuvered.

Zed shook his head. "Not yet. We have more important things to consider, before we go joy-riding all over hill 'n dale."

Disappointed, I asked, "What are we considering?"

"The other types of fuel cells. There are five in all, each with its own applications. Even though the PEM may be the only type of fuel cell you will ever encounter, you should still know about the others."

"Why?" I asked, staring longingly at Zed's little yellow car.

"Because fuel cells represent a major shift in how we think about power generation. Besides using fuel cells in our cars, we can use them to

Zed sez

Fuel cells represent a major shift in how we think about power generation. Besides using fuel cells in our cars, we can use them to supply power to remote towns and villages. They can also be used as steady, backup power for hospitals and other institutions where power outages or voltage fluctuations are not acceptable.

supply power to remote towns and villages where it's impractical to build power generating plants, or string countless miles of inefficient power lines. They can also be used as steady, backup power for hospitals and other institutions where power outages or voltage fluctuations are not acceptable."

It seemed there was a flaw in his reasoning. Considering the expense of manufacturing, transporting and storing hydrogen, it hardly seemed practical to use fuel cells for electricity generation on such a large scale. What possible benefit could there be? So I asked.

"Who said anything about hydrogen?" Zed answered, looking puzzled.

"Well, I just thought that..."

"All fuel cells run on pure hydrogen? You're right, of course. They do. But high-temperature fuel cells—the big, hot, beastie boys that operate at over 600 degrees Celsius—can also use natural gas, syngas, or even ethanol and methanol. And use it more efficiently than conventional power plants."

"That's pretty hot, Zed," I interjected. "At that temperature it's no wonder they can be made to run on about anything."

"True. And because they operate at such high temperatures, the 'waste' heat they produce can be used to produce steam for heating or industrial processes."

Industrial processes? That caught my attention. "How big are these things, anyway?"

"As big as you want them," he answered. "They can be sized to produce electricity in the megawatt range. And several can be linked together to produce as much power as you need."

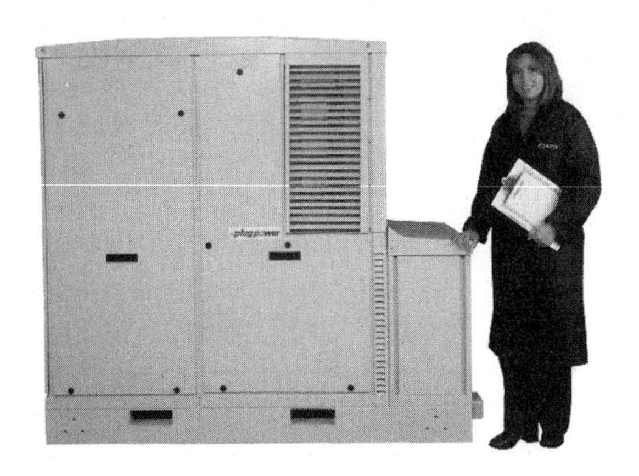

Plug Power's installation of grid-connected PEM fuel cell systems provide electricity to 1.1 million customers throughout Long Island Power Authority's service area in New York. *Courtesy of Plug Power Inc.*

Zed sez

Hundreds of these phosphoric acid fuel cells have already been installed around the world, in units producing 200 kilowatts each. So, at least to this point, phosphoric acid fuel cells have probably made more electricity than all other fuel cells combined.

"What do these things use as electrolyte, Zed? Obviously not a sheet of plastic."

He laughed. "No, I'm afraid that wouldn't do. Actually, there are two types of high-temperature fuel cells. Molten carbonate fuel cells use lithium and potassium carbonate as the electrolyte, and are the cooler of the two types, operating at around 650 degrees Celsius. Solid oxide fuel cells, on the other hand, use what's called 'Nernst-Mass' as an electrolyte. This is a mixture of zirconia and yttria—an oxide of the rare-earth element, yttrium. These cells operate at around 1,000 degrees Celsius, which falls just between the melting points of silver and gold."

"Nothing you'd want to put in your car, in other words."

"Your car? No, but you could put one in your house. Solid oxide fuel cells can be sized down to as small as two kilowatts—about the amount of power it takes to run a hair dryer and coffee maker simultaneously."

"Wonderful," I replied, eyeing the sporty yellow two-seater and hoping the conversation would veer back in that direction.

Zed didn't seem to notice my impatience. "The phosphoric acid fuel cell, on the other hand, is a medium-temperature fuel cell, and another type that can be used in fairly large installations, though not usually as large as the high-temperature boys. Although it runs cooler—only at about 220 degrees Celsius—and therefore needs pure hydrogen, the hydrogen can be supplied from a built-in natural-gas reformer. Hundreds of these have already been installed around the world, in units producing 200 kilowatts each. So, at least to this point, phosphoric acid fuel cells have probably made more electricity than all other fuel cells combined."

Four down, one to go. "And what's the last one?" I asked, hoping to look interested. "Besides the PEM fuel cell, that is."

Zed replied, "That would be alkaline fuel cells. They use sodium or potassium hydroxide as the electrolyte. Like PEM fuel cells, alkaline fuel cells operate at fairly low temperatures—between 60 and 120 degrees Celsius—and are suited for applications in the 5 kilowatt range. Besides that, they're reliable and can be cheaply made."

Suddenly I was getting interested, again. "So what's the problem?" I asked, "why is everyone so high on PEM fuel cells if alkaline fuel cells do the same thing for less money?"

Zed answered, "They have one major flaw—an Achilles heel, you might say. The alkaline electrolyte absorbs CO_2 from the air. This eventually dilutes it, making it less efficient at carrying a charge. So, unless you can feed it pure hydrogen on one end, and 'scrub' the air clean of CO_2 on the other end, you'll quickly run into problems."

"Sounds serious. Why would anyone want to use them?"

"Oh, they have their uses. Alkaline fuel cells were used in the Apollo Space Program to help us get to the moon, and they still supply the electrical power for the space shuttle. But unless someone figures out a way

5 Types of Fuel Cells

	Fuel Cell	Operating Temperature	Efficiency	Power	Uses
1.	**PEMFC** *Proton Exchange Membrane*	50°-100° C	35 - 45%	5 - 250 kW	CHP transportation portable power
2.	**AFC** *Alkaline Fuel Cell*	60°-120° C	35 - 55%	< 5 kW	military/space
3.	**PAFC** *Phosphoric Acid*	220° C +/-	> 40%	200 kW	CHP transportation
	Large Scale Stationary Power Generation				
4.	**MCFC** *Molten Carbonate*	650° C +/-	> 50%	200 kW - MW	CHP
5.	**SOFC** *Solid Oxide*	1000° C +/-	> 50%	2 kW - MW	CHP

CHP = Combined Heat and Power Generation (electric utility) kW = Kilowatt MW = Mega Watt

to overcome the CO_2 problem, they'll never compete with PEM fuel cells in the mass market."

With transparent intent, I said, "Which brings us back to your delicious little car, here..."

"...And the fact that you still don't know how a fuel cell works."

"Yeah, well..."

"This car can wait," he admonished, in his best pedantic tone, "until you have at least a cursory understanding of how it works. Come on, I've got something to show you."

At that point, the little alarm went off in my head. He was gearing up for a hands-on demonstration, but there was nothing in sight. Where was the demonstration? Wasserstoff World, that's where. It was the last attraction, a weird-looking edifice off in the distance, behind the Wasser Kleaver electrolyzer and to the left of the Pickett Meg-3 wind turbine. That just had to be it. To top it off, it was in a country Zed refused to talk about—probably hostile to everyone but eccentric Druids—and accessible only through a means of travel as spooky as a time machine. I wasn't going, no way, no...

Zed lowered the body back down over the car and secured it. Then, innocently, he said, "It's over here in the greenhouse."

"...Huh? The greenhouse, you say?"

"Of course. Hurry up, let's go."

I followed him toward the greenhouse, happy to be getting farther away from the dimensional compressor room with every step. Zed opened the greenhouse door, I naively stepped in, and—*Voilá*—I was standing in the middle of nowhere with nothing but sun-burnt sand beneath my feet.

"*You tricked me*!" I cried, the second Zed materialized next to me.

Indignantly, he retorted, "I did no such thing. I told you it was in the greenhouse, and it was."

Zed sez

Alkaline fuel cells were used in the Apollo Space Program to help us get to the moon, and they still supply the electrical power for the space shuttle.

"Yeah, but I thought..."

"I can't help what you thought. I decided to put the dimensional compressor in the greenhouse, hoping the magnetic flux waves from the spinning terra-globe in the back room would smooth out the eddies in the hyper-dimensional field equations. It seems to work like a charm," he said, tugging on the long, white locks that hung compliantly over his shoulders. "Anyway, here we are, so if you're through whining, maybe we could get on with it?"

I just shook my head. "Whatever. I guess as long as we're here."

"Good. The latest Wasserstoff World attraction is right behind you. I call it 'The Electron Odyssey'."

"Catchy," I said, as I turned to see what Zed was talking about. I'd have thought it would be a much bigger version of what was in the car—something like a building-sized suitcase—but this thing was several stories high and no more than a few feet thick; flatter than a pancake, relatively speaking. I could see an electric motor mounted on top of the whole apparatus.

"That's it?"

"That's one cell. It's the basic building block of a stack of many cells, which is what fuel cells really are...as I explained earlier, while you were making goo-goo eyes at the car."

"It's a sexy car," I said in my defense.

"So it is. And the sooner we get through this, the quicker you can drive it."

"I'm all ears," I assured him.

"Good, now pay attention. You will notice seven distinct layers here. Do you see them?"

I studied the side of the cell, and said, "Yeah, if you count the two skinny layers on either side of the middle layer."

Zed said, "Yes, well, that middle layer is the proton-exchange membrane, and the two 'skinny' layers are the catalysts. Presumably, they're rather important, so we really should count them, don't you think?"

I scowled. "You don't have to get sarcastic about it, you know. Anyway, I had that much figured out. What I don't know is what to call those two thick layers attached to the catalysts."

"Those are the gas diffusion electrode layers. Together with the catalyst layers, they form the anode and the cathode. They're porous to allow for the movement gases—hydrogen on the anode side, oxygen on the cathode side—and conductive, to facilitate the movement of electrons. They're made from carbon cloth."

That finished the inside of the sandwich. Now for the bread. I asked, "And the metallic-looking plates on either end? The ones with the channels cut through them? What do they do?"

"They'e the flow-field plates. They conduct electrons from the electrodes on the anode side, ferry them to the load—the motor on top, in this case—and back to the cathode to complete the circuit. The channels you see are called flow fields. It's their job to disperse the gases evenly from one side of the cell to the other."

Now that Zed had explained all the layers of his oversized PEM fuel cell, I expected it to light up, or become transparent, or something. But it just sat there. Puzzled, I asked, "What happens now?"

"That depends," he answered coyly.

"On what?"

"On if you want to be the electron or the proton."

I smiled nervously. This was beginning to sound like a wild ride. "Which would you suggest?"

He said, "Since this is your first time, I would think you'd be more comfortable as the proton."

"Proton it is. Let's do it," I said, blissfully ignorant of what I'd just agreed to.

Zed led me up a stair to a platform that rested under an opening in the flow field. He stood close to me and winked reassuringly, as my body began to tingle all over and I began to...shrink? I looked at Zed as if to ask what was happening, but he had grown much smaller than me and had drifted a great distance away, where he began to orbit around me like a small moon.

Quickly we were whisked away like a dust particle in the breeze and sent careening through the channels of the flow field—which now seemed miles across—until we were drawn into the carbon-cloth electrode layer. Each fiber, made up of countless layers of neatly stacked carbon atoms, was far too big to see in its entirety, and the spaces between them were like vast canyons. I felt like a bird in flight through a planet-sized rag.

Soon we reached the catalytic layer and, though Zed was so far away and spinning so fast I could scarcely see him, I heard his voice inside my head, directing my attention. "The catalytic layer is made up of porous carbon and infinitesimal particles of platinum," he whispered from between my ears. "The platinum is the catalyst that makes this all possible. Any second now, it's going to grab us and then we'll go our separate ways. I'll follow the electrical

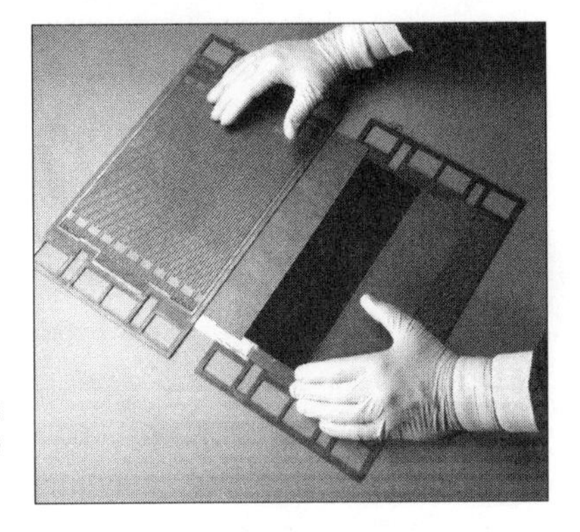

Two bi-polar (flow field) plates and membrane electrode assembly (PEM and catalysts). *Photo © General Motors.*

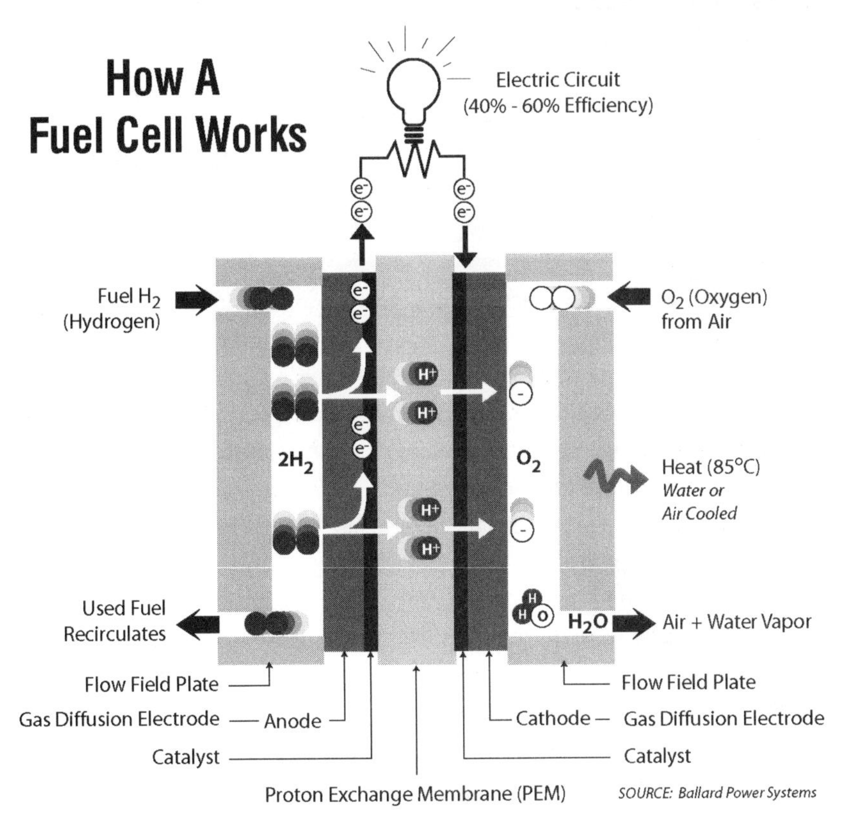

potential in the circuit that goes outside of the fuel cell to power the electric motor. You'll go right through the heart of the thing."

Like a fly in a spider's web, it happened. For the briefest moment I felt as though Zed and I had somehow become part of a platinum alloy, but then he was quickly pulled away from the force field that had held us together and sent back into the carbon cloth electrode layer.

Though far from my sight, I could still hear him inside my head. "Whooo-eee!" he screamed, with childish delight, "this is always such a *rush*!"

As soon as Zed was spirited away, the platinum particle—a beautiful nucleon-laden metal—released me, and I caught an unlikely ride on what had to be a water molecule. The familiar, Mickey Mouse-like agglomeration of hydrogen and oxygen quickly took me into the heart of the fuel cell: the proton exchange membrane.

Again I heard Zed: "By now you've probably caught your ride. Don't be alarmed. The only way you'll get through the membrane is on the back of that water molecule, so don't let go—" he laughed, devilishly "—not that you could if you wanted to."

It was true—I was stuck. Very quickly, however, things changed, for the cathode (identical to the anode, I quickly saw) loomed straight ahead. A diatomic oxygen molecule lingered on the surface of a platinum particle, as if waiting for me. But where was Zed?

"Hold on, lad!" he screamed, this time from a great distance, "I'm on my way!"

Everything happened so quickly after that it's difficult to describe. In a rush, three protons I hadn't seen before came up from behind me and bounded off their water ferries onto the oxygen atoms. They were quickly joined by three electrons, and me and Zed, who looked like he's just ridden a lightning bolt to the moon and back. Suddenly, Zed and I had become part of a pair of water molecules.

But not for long, fortunately, for no sooner had the transformation occurred than we were on the platform on the other side of the fuel cell—full-sized, breathing, and soaking wet.

"What a ride!" I huffed, between gasps.

Zed replied, "And you took the easy route." I looked at him, then, and I'd never seen him so frazzled. "I think I'll replace that motor with a lightbulb," he said, shaking his head. "It's got to be easier than spinning through those copper windings at 12,000 rpm."

Back at the Wasserstoff Farm, the little yellow car looked positively tame after what we'd just been through. But no less fun. "Can we drive it, now?" I asked, my voice sounded tired, even to me, but still bubbling with anticipation.

"Sure, why not? I guess we're dry enough."

Zed sat down into the passenger seat as I slid in behind the wheel. Like Zed's custom golf cart, the little roadster took off, quick, smooth and quiet as a whisper. And though it didn't have the G-force effect of his souped-up muscle car, it had more pep than I would have thought. Best of all, it didn't have the feel of a car; it was more like riding on a low cloud, so silent and smooth, and free of vibration.

I heard Zed talking as I drove...

The instant acceleration is due to a high-voltage battery that kicks in when you hit the pedal...

The battery is charged by a regenerative braking system...

The 3-phase induction motor can reach 67 kilowatts of power, at up to 91 percent efficiency...

...but I wasn't listening. The top was down, the wind blowing fresh and cool against my face, and I was in control of a magic carpet.

It really can't get any better than this, I thought. What a wonderful technology.

technistoff - 17

For those who would claim that Zed monkeyed with the laws of physics to make his *Electron Odyssey* work, I would be forced to agree, on several counts. The idea that two humans could suddenly be so closely endowed with the physical attributes of a hydrogen molecule that they could fake their way through a fuel cell is patently absurd. The relative speed would have been terrifying and the collisions between particles bone-crushing; hardly the leisurely ride it appeared to be.

However, were Zed to have made the *Electron Odyssey* accurate on an atomic level, everything would have happened so fast there would hardly be a story to tell, so blind is the human sensory apparatus to events in the microsecond time frame. Besides that, Zed is—admittedly—a humble sorcerer trying to turn a buck on a risky theme park that's obviously not optimally sited.

On those considerations I, for one, am willing to cut him a little slack.

That being said, as graphically profound—and as riotously fun—as it is to zip through the heart of a fuel cell on the back of a water molecule, it's difficult to understand what's happening when you are right in the middle of it. I was left with a question or two about how the whole process worked.

how fuel cells work

Why was Zed (the electron) compelled to go up and out through the electrode, while I (the proton) was not? This all has to do with what is called "electrical potential," which loosely means "a whole bunch of negatively-charged electrons piling up in one place, trying to find a path to a gang of positively-charged ions at the far end of the circuit." The ions, of course, are protons. As the protons take the shortcut through the electrolyte, the electrons are forced to take the long way around. Before the electrons finally make it to the cathode (where the protons hang out with the O_2, waiting for water to happen), they may be called upon to perform any number of unsavory acts—running motors, lights, radios, horns, etc.—which they do willingly, since they are so intent on joining back up with their long lost protons.

Why does the proton exchange membrane permit the passage of protons, but not electrons? Because of negatively-charged, sulfuric-acid side chains attached to the polymer matrix, the entire membrane carries a net negative charge. It repels electrons, while attracting protons. To keep the sulfuric acid (SO_3^-) ions in the side chains from holding on too tightly to the positive hydrogen ions, water is needed to attract the protons and carry them through the membrane, from one SO_3^- ion to another.

What's in it for oxygen? Not much, from an ionic standpoint. While the electrons zip away like juvenile delinquents with the old man's car, and protons seem to be on a quest from which they can scarcely be deterred, the O_2 molecules just sort of hang around for lack of anything better to do. If it weren't for the Platinum Escort Service (and a lot of pressure from the onboard compressor), which keeps everything together long enough to make water, a fuel cell would be a real uninteresting place.

After my ride through the *Electron Odyssey*, I read dozens of explanations on how PEM fuel cells worked, outside of theme parks. The very best was at the Los Alamos National Laboratory website at: *http://education.lanl.gov/resources/fuel cells/fuelcells.pdf*. It's thorough, yet understandable; science as it should be.

remaining challenges

How close are we to driving fuel cell-powered cars? PEM fuel cells are today sufficiently energetic and compact to meet anyone's standards for power and efficiency. They are, however, far from being perfected. The cost of the platinum needed for the electrodes is too high. New electrode materials are needed to bring down the cost of fuel cells and to improve the dismal efficiency of the cathode (oxygen) reaction, which proceeds several times slower that the anode (hydrogen) reaction. PEM fuel cells also require hydrogen of very high purity, since carbon monoxide (as may be

Because there are no harmful exhaust gases, fuel cell cars are often called zero-emission vehicles.

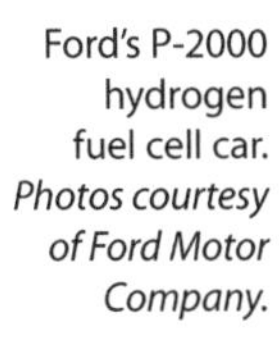

Ford's P-2000 hydrogen fuel cell car. *Photos courtesy of Ford Motor Company.*

present in hydrogen made from syngas or natural gas reformate) attaches to the platinum electrode, and doesn't let go.

Research is currently being conducted to find ways to optimize the surface area of a given amount of platinum, as well as ways to remove CO from hydrogen. But finding a way to entice oxygen to be a little more liable to react with hydrogen at the cathode is an ongoing problem.

fuel cell cars in the works

Nonetheless, virtually every major automaker has at least one fuel cell car in the works. For a comprehensive list, see: *www.fuelcells.org/fct/carchart.pdf* at the Fuel Cells 2000 website. The Ford Focus FCV-Hybrid is an excellent example of how seriously Detroit is taking the prospects for a hydrogen economy. Though still several years away from mass production, this four-door sedan shows real sophistication for a prototype car. It gets its power—up to 87 horsepower—from a Ballard Mark 902 fuel cell stack. For quick response times and extra acceleration, a high-voltage Sanyo nickel-metal hydride battery (recharged, in part, by a regenerative braking system) can kick in an additional 25 horsepower. Like Zed's little yellow sports car, the Focus FCV-Hybrid has a single-speed transaxle, and is capable of speeds over 80 mph *[128 km/hr]*, which it can reach in about 25 seconds. It runs on hydrogen supplied by a single 5,000 psi tank, has a driving range of 160–200 miles *[260–320 km]*, and an average fuel economy equivalent to 50 miles per gallon *[21 km per liter]* of gasoline.

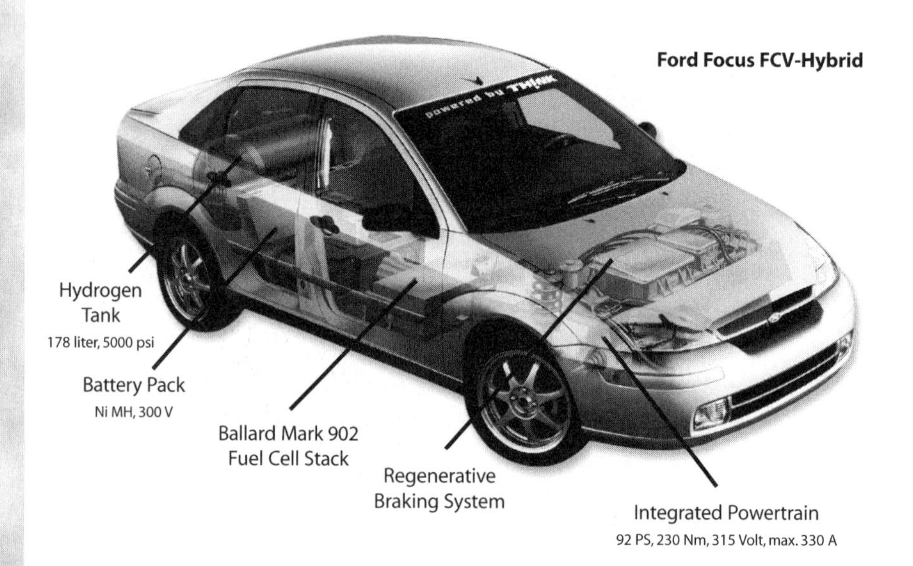

An excellent article by Stefan Geiger, on the design and performance of the Focus FCV-Hybrid car can be found on the Fuel Cell Today website at: *www.fuelcelltoday.com/FuelCellToday/FCTFiles/FCTArticleFiles/Article_516_FordFocusFCV.pdf.* The car's specifications can be downloaded at the Ford Motor Company website at: *www.ford.com/en/vehicles/specialtyVehicles/environmental/fuelCell/focusFCVHybrid.htm.*

GM's Hy-wire car

As you might imagine, General Motors is not standing idly by as the race heats up. GM's Hy-wire concept car is a radical new chassis design that demonstrates fuel cell and by-wire technology in an innovative way. According to Larry Burns, GM's vice president of R&D and Planning, "All of the touring sedan's propulsion and control systems are contained within an 11-inch-thick skateboard-like chassis, maximizing the interior space for five occupants and their cargo." This includes a trio of 5,000 psi (350 bar) hydrogen fuel tanks and a hefty 94-kilowatt PEM fuel-cell stack. "There is no engine to see over,

The Hy-wire has no internal combustion engine, instrument panel, brake or accelerator pedals—but it does have ample power supplied by a GM fuel cell that runs on hydrogen. *Photos © General Motors.*

General Motors' Hy-wire Car

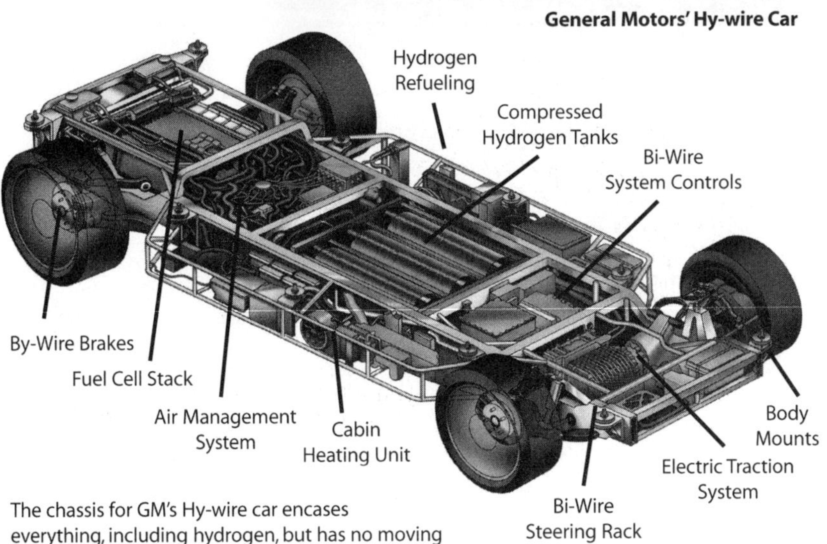

The chassis for GM's Hy-wire car encases everything, including hydrogen, but has no moving or mechanical parts. Bodies (tops) are interchangeable.

no pedals to operate—merely a single unit called the driver control unit that is easily set to either a left or right driving position." Braking and acceleration are controlled by twisting the handgrips, which glide up and down for steering. This means that all the driver's controls are electronically activated, unlike standard cars where the brake, accelerator and steering linkages are—at least in part—mechanically operated by the driver. The Hy-wire car is likely to open a new world of possibilities for vehicle designers to explore, due in part to its simplified engineering and reduced number of parts.

In their quest to make fuel-cell vehicles a reality, GM has teamed up with some impressive players, including Quantum Fuel Systems Technologies Worldwide, for high-pressure hydrogen storage systems; Hydrogenics, for fuel cell product development; BMW, for development of refueling devices for liquid hydrogen; and Toyota, for collaborative development of fuel-cell technologies. Seems these folks are serious. To learn more about what the GM fuel-cell team has been up to, visit: *http://www.gm.com/company/gmability/adv_tech/ 400_fcv/index.html.*

methanol fuel cells

What about methanol? It's a liquid fuel that's easy to transport and store, and can be handled with the same relative ease as gasoline. Moreover, because it can be mixed with water, the problem of keeping the proton exchange membrane hydrated (without flooding it) is easily solved. But will it ever be a practical fuel for fuel cell cars? The answer is: probably, someday.

Current attempts to use methanol directly in a PEM fuel cell (i.e. a direct methanol fuel cell) suffer from two problems: low reactivity at the anode, and methanol crossover through the membrane to the cathode. Research in this area is focused on finding more reactive compounds for the anode, since the amount of platinum needed for reasonable reaction speeds is several times greater than what is needed for a pure hydrogen fuel cell. They are also looking into developing different membrane materials impermeable to methanol.

This doesn't mean methanol is out of the picture. Several car manufacturers are working with prototypes that use standard PEM hydrogen fuel cells, and onboard reformers to extract and purify the hydrogen from methanol. In fact, many hydrocarbons can be reformed in this way, including gasoline. The reformer, however, adds unwanted complexity to the system and cuts into fuel economy, since it requires a significant fraction of the fuel to operate. For this reason, it's doubtful methanol will be used extensively in hydrogen fuel cells until it has a fuel cell it can call its own.

Research on fuel cells that burn straight ethanol is even further behind since, until recently, the distillation process used to make it relied on a very limited list of feedstocks. With the introduction of new ways to effectively distill ethanol from a much wider range of biomass materials, this is probably due to change. At any rate, once an efficient direct methanol fuel cell emerges on the scene, a direct ethanol fuel cell will probably not be far behind.

ethanol fuel cells

At this stage of the game, fuel cells are more a curiosity than a common means of producing electrical energy. Most of us have never seen or touched a fuel cell, and even fewer have driven a vehicle, or lived in a home, powered by one. We should all be prepared for that to change. By the end of the first decade of the twenty-first century, fuel cell cars will appear on our roads, as will the refueling stations needed to keep them going (though it will probably be quite some time before they're as common as the corner gas station). Home-based wind and solar-electric systems that now use batteries for energy storage may one day use electrolyzers and fuel cells in their place. As solar cells become cheaper and more efficient, their use will be proportionately more widespread, leading perhaps to an ever greater presence of fuel cells in our homes.

Beyond our homes and our cars, fuel cells will begin showing up wherever electricity is needed, from laptop computers and cell phones to small 1 to 2 kilowatt units designed to replace the noisy gas generators used for camping.

Right: The AirGen™ is the world's first portable fuel cell generator for indoor operation, powered by Ballard's leading fuel cell technology. It uses oxygen from air and hydrogen fuel to create electricity; by-products of the reaction are heat and water. *Photo courtesy of Ballard Power Systems.*

Multi-megawatt installations such as the postal sorting facility in Anchorage, Alaska, that uses five PC25 200-kilowatt phosphoric acid fuel cell stacks (manufactured by UTC Fuel Cells) to supply all of its electricity as well as a good portion of its heat, will become more common.

beyond automobiles

To learn more about the science of fuel cells and their applications, I would highly recommend the *Fuel Cell Technology Handbook*, edited (and partly written) by Gregor Hoogers. It's rigorous, but it's all there: how each type of fuel cell works, in theory and in practice, and what it's going to take to make them work better; details about specific installations (such as the postal facility mentioned above); which fuel cell works best for which application, and why; and a thorough discussion of the current status of fuel cell vehicles.

Fuel cells for our cars and our homes will almost certainly be PEM fuel cells, and many of them will be manufactured by Ballard Power Systems of Burnaby, British Columbia. The Ballard website (*www.ballard.com*) has a wealth of information, including a short movie explaining how a fuel cells works to power a car.

It's not Wasserstoff World, mind you, but it's not bad.

FUEL CELLS TO GO

fuel cells for laptops and cell phones

Long before you ever get the chance to cruise smoothly and quietly down the road in a fuel cell powered car, or generate clean, dependable power from a fuel cell stack tucked neatly away in your basement, you'll be using small, relatively inexpensive fuel cells to power your cell phone and laptop computer.

A number of companies, including Millennium Cell, Neah Power Systems and Toshiba, are working on portable fuel cells to power mobile electronics, and they expect their new power packs to be ubiquitous by the end of the decade. All use some variation of the proton exchange membrane (PEM), and most will be powered by methanol or ethanol. Millennium Cell is working on a fuel cell powered instead by sodium borohydride, a clever hydrogen carrier discussed in Chapter 15, *Stuffing Wasserstoff*.

The question is, why would we *want* fuel cells, when we've gotten by so well for so long on batteries? Simply put, the sophistication—and therefore the power demands—of our cell phones and laptops is outpacing the battery technology used to power them. Cell phones with color displays are doing double-duty as digital cameras, and today's laptops, which sport 15- and 16-inch screens and super-powerful processors, are more watt-hungry than their predecessors.

portable recharging units

The first wave of small fuel cells will probably not replace the batteries in these devices directly, but will instead serve as portable recharging units. Inexpensive, self-contained, recyclable fuel packs will provide the power, and will be as easy to change out as batteries in a small flashlight.

How is it that methanol and ethanol can be used in small fuel cells, while in larger ones they suffer from the problem of membrane-crossover and lack of reactivity at the anode? The answer lies in the fact that these fuels are used in very dilute concentrations in small fuel cells, while specialized electrodes and membranes allow for optimal fuel utilization.

For more information about portable fuel cells, read the November 2003 *Computer* magazine article by Linda Dailey Paulson, "Will Fuel Cells Replace Batteries in Mobile Devices?" (available for download on the Neah Power Systems website at: *http://www.neahpower.com/news/newsfiles/IEEEComputer.pdf*), or read "Micro-Tech, Mega Potential," by Rona Fried, Ph.D., in the May/June 2004 issue of *Solar Today*.

HOME-GROWN HYDROGEN

Will it ever be possible to create hydrogen at home, using only sunlight and wind, and then use the hydrogen to run electrical appliances, heat the home, and maybe even as fuel for the family car? The answer is "yes," but not for a few years.

fuel cells to provide home electricity

Current home-based renewable energy systems use solar-electric (photovoltaic or PV) modules and small wind turbines to create DC electricity, which is stored in large banks of batteries. When electricity is needed, the low voltage DC (usually between 12 and 48 volts) travels from the batteries through a power inverter, where it magically becomes 120-volt AC house current. The inherently low efficiencies of PV modules and wind turbines notwithstanding, the solar/wind-to-electrical-load-efficiencies for this type of system are in the range of 80 to 90 percent.

In a hydrogen-based system, by contrast, the batteries would be replaced by an electrolyzer, a hydrogen storage system, and a fuel cell stack. By going from electricity to hydrogen, then back to electricity, over 65 percent of the original energy is wasted using current technology, leaving you with a mere 35 percent of the power you started with. So why would you even bother?

A couple of reasons. First, a fuel cell can deliver power faster than batteries, and will have a longer useful life. Also, unlike a battery that must keep 50 percent of its power in reserve to avoid damage, a fuel cell will deliver full power as long as there is hydrogen to fuel it.

hydrogen also for heating and cooking

In addition to electricity, hydrogen can be used for home heating and cooking, and can even be used as car fuel. So how do you make the process less wasteful? By using a high-temperature fuel cell (such as a solid oxide fuel cell), the excess heat can be used for water or home heating, thus increasing the efficiency.

Even if electrolyzers could someday reach 90 percent efficiency, and fuel cells could have a combined heating and power efficiency of 70 or 80 percent, that's still less than batteries. But what if PV modules were twice as efficient and half as costly as they are now? Then suddenly deluxe, home hydrogen-based systems could become even more affordable than today's battery-based system.

Get ready—it's coming.

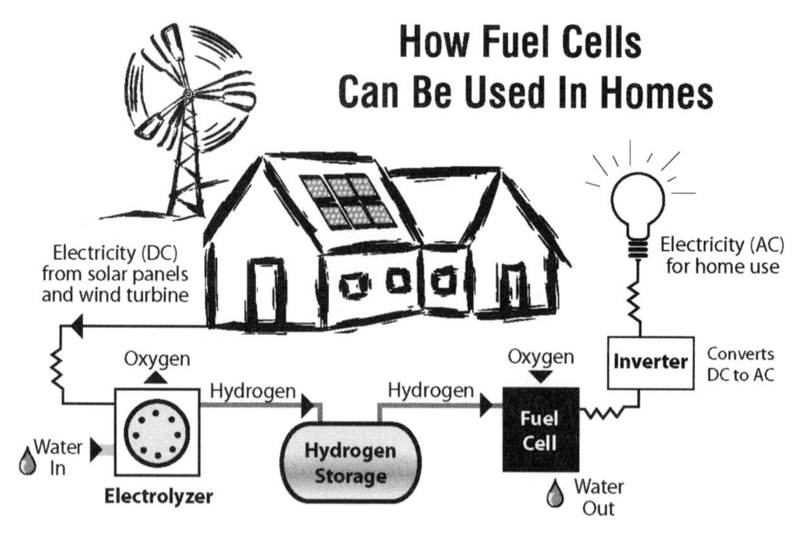

Energy from the sun and wind can be used to create hydrogen by electroylsis. The hydrogen is stored until electricity is needed. A fuel cell then takes the stored hydrogen and adds oxygen from the air to create electricity. Heat created in the fuel cell can be captured for home heating. Water is the only by-product.

chapter 18

Night Vision: A Peek Into Tomorrow

It was still dark when Zed woke me up in the early morning hours of what was to be my final day at the Wasserstoff Farm. Instead of slamming the door, bouncing on the bed, or plying me with coffee, he simply whispered gently in my ear, "We're going to take a little trip, but I want you to keep your eyes closed until I say to open them. Can you do that?"

I was still half asleep when I answered, "Huh? Sure Zed, but—"

"No 'buts'. Keep them closed," he insisted, speaking in hushed tones.

"Alright. I will," I agreed, feeling my heart rise into my throat. Anticipation or trepidation? With Zed, they were often two grapes from the same vine.

I felt his hand grasp my arm lightly and I rose from the bed. I had the sensation of being as light as a feather, or—and this was more the truth of it—a spirit. As I think about it now, it seems odd that I felt no fear as Zed took me up, higher and higher (certainly through the roof of my cottage, passing from ceiling to roof tile with the ease of a rose petal rising on the breeze through a layer of mist), but I felt nothing but a deep sense of wonder. In the past few days my notion of impossible—that impenetrable wall that always seemed to thrust itself skyward on the path just in front of me—had grown small and transparent, and I knew I liked it that way.

After a short time, though it may have been an hour or a day, so inconsequential did the ticks of the clock seem, Zed released my arm and instructed me to open my eyes. Since I'd had no sensation of traveling in any direction but up, I was more than a little surprised to not be looking at the roof of the cottage and the other buildings that surrounded it. In fact, there was nothing below us that seemed in the least familiar.

"What do you see?" Zed asked.

"Chaos," I answered. "Everything's in a jumble."

"Not chaos, really. It's just the future, unformed. It's waiting to be molded into whatever shape the collective mind of mankind deems prudent or expedient. Or—and at this I shudder—inevitable. Remember that nothing is inevitable, lad, at least not until everyone gives up and allows inertia to rule the coming day."

"So how do we make things change, Zed? Right now it's like looking down through the funnel of a tornado."

He answered, "What do you believe will happen?"

Feeling a little ashamed, I replied, "I guess I haven't thought about it. I've been too busy learning *how* it all works to give any thought for how it all works *together*."

"Ah, yes...a not uncommon problem. Well, here's your chance, if you dare to dream a little."

Floating, as we were, high over a vast swirling sea of possibilities, it seemed there was no limit to what could be done with enough resoluteness. I closed my eyes and concentrated (with all my limited abilities in that regard) on what seemed a doable, even likely, beginning. When I felt I had reached my limits, I heard Zed say, "Good, lad. You may have a talent for this sort of thing."

Opening my eyes, I looked down and saw dozens of vehicles—buses, taxis and corporate fleet cars and trucks—all running on hydrogen, silently and pollution-free over city streets and rural highways. Soon, their number grew into hundreds, then thousands. Quickly, these were joined by private cars; a few at first, but shortly they became more numerous than the public and fleet vehicles. In unison, filling stations with huge storage tanks of compressed hydrogen gas appeared and quickly grew in number. "Where are all the other cars, Zed? You know, the normal ones?"

"Oh, they're there. Gasoline-electric hybrids, battery-electric cars, plus a few of the good old gas guzzlers. It's quite a place."

I wondered where the hydrogen was coming from, and then I saw: natural gas reformers, mostly, but also from grid-connected solar power towers, solar troughs and wind farms. It was exciting. "Where is this, Zed? Where is this happening?"

"It's *your* vision, lad. Look closely."

I picked out a bus traveling along a rural byway and focused intently on the license plate. "Of course," I smiled. "California."

"It all has to start somewhere," Zed interjected. "As with home-based solar and wind generated electricity, the initiative often starts at the state level. It's better that way, really—a little healthy competition among states is a good thing."

"What happens now?" I asked.

Zed graced me with a rare chuckle. "You tell me. I'm just your escort."

"Well, it would seems to me that other states would follow. Everyone's going to try and outdo the leader. This will inevitably lead to mistakes and oversights, but they'll turn out to be just bumps in the road, because once the ball gets rolling, nothing is going to stop it."

"You may be right. I hope so," Zed interjected with a note of caution.

Filled now with an energized enthusiasm, I exclaimed, "Believe me, Zed. I can see it! What is today merely an industry of promises will shortly be a juggernaut. No one will be able to ignore the potential to make money, and the pace of research—both public and private—will reach a feverish level. I see everyone getting into the act. The coal interests, who profess to have the technology in hand to make emission-free electricity and hydrogen from the dirtiest of fuels, will have to clean up to stay in the game. And they will, because—for the short term, at least—it simply won't be possible to make enough hydrogen without using fossil fuels to produce the electricity, or the heat, to separate hydrogen from water. But the old, dirty coal-burning plant's day is done—it will be gasification from here on out, and then only until renewables are able to stand alone."

"And nuclear? Do you see a place for nuclear?" Zed asked.

I looked down, and wrinkled my brow. "It's not clear, Zed. It's there, and then it isn't. It fades in and out."

Reassuringly, he said, "Don't worry, lad. Perhaps there are too many variables; too many unanswered questions. It will become clear in time, as all things eventually do."

What was unambiguously clear was the role played by renewables in the hydrogen economy. As Zed and I floated ever higher and my view broadened, I could see wind farms springing up all across the Midwest and high plains, and vast solar installations—from power towers to photoelectrochemical arrays—dotting the landscape across the Southwest, from Texas to California. Wherever there was enough wind or sunshine, there was a facility designed to harness the energy.

Nor was the surge in renewables limited to massive commercial operations. A new generation of efficient and inexpensive photovoltaic products appeared as roofing and siding materials on millions of homes throughout the country, providing so much home-based electricity that dozens of old, inefficient coal-burning power plants were decommissioned.

In rural areas, thick stands of switchgrass, poplar trees and other fast growing biomass species sprang up in idle farmland in every state, and with them distillation and gasification plants producing methanol and ethanol. Cars appeared that ran on one or the other—or both; internal combustion engines and fuel cells, alike. And hydrogen?

I asked, "Zed, is it possible for a fuel cell to run on both alcohol and hydrogen, or is this just wishful thinking?"

"No theoretical reason why not," he said.

"What about internal combustion engines?"

"Eminently doable," he replied.

I smiled at that. Perhaps the competition between hydrogen purists and alcohol advocates would, in the end, become a partnership. With proper processing, I recalled, the burning of alcohol could actually reduce the amount of carbon dioxide in the air, and I could see it happening before my eyes. The skies were becoming clear, first over the countryside, then over the vast metropolitan areas. It wasn't just carbon dioxide, of course; it was everything. With no source of replenishment, the soot and smog—mostly particulate carbon and nitrogen oxides—was dissipating. For the first time in decades, people in the cities were coming out of their homes at night to gaze at the stars.

I took a deep, satisfying breath, and said, hopefully, "Is this it, Zed? Is this really the way it will be? Or is it only a dream—a vision I cooked up in my imagination?"

"It depends, lad, on how many others see what you just saw. There's nothing in your vision that can't be done, if enough people want it to happen. But if research isn't adequately funded, or short-sighted politicians only offer the usual "someday" lip service? Well then..."

Far below, on the dreamy landscape of future possibilities, I saw a disheartening scene. A few buses and corporate vehicles running on hydrogen, while all other modes of transportation were gasoline- and diesel-powered, with a few token hybrids. Coal plants continued to spew out carbon dioxide and noxious sulfur compounds, while teams of engineers lamented the "tremendous capital costs" of cleaning them up. The acres dedicated to biomass crops were but a fraction of the acres that remained fallow. A few wind farms sprang up here and there, and enough solar installations to make a dent—a pitifully small one—in the growing energy demands of a country numbed by its own excesses. And all the while the air grew murkier and the people more hardened by their discontent.

It was a sad scene to see. I felt deflated.

"Zed? Please tell me it's not real."

He patted me reassuringly on the back, and I bobbed in the mist like a cork in a pond. He said, "It isn't, lad. At least, no more real than what you saw at first. Both are possible, perhaps even equally probable. But the two possibilities are not, as you may think, separated by cost considerations, or technological hurdles, but by attitude—a mere state of mind. If the people of a society want something bad enough, they can have it. We've always managed to find the money and the technological resources to protect our country from any threat whenever the danger was apparent. And perhaps that's the key: if we truly understand we *need* it—if we acknowledge that, ultimately, it's either the hydrogen economy or a return to the stone age—then we can have it. If not, then nothing you've learned at the Wasserstoff Farm matters."

I asked, "What's it going to take, Zed, to make the first vision real? How do you change the attitudes of a quarter billion people?"

He laughed. "You humor them, lad. And then you confound them with leadership."

Before I could ask what he meant, he was gone. I felt myself gently gliding downward, and as I did so I felt my consciousness slowly slipping away...

As the early morning light began to filter through the cottage's windows, Zed awakened me for the second time. Or was the first time only a dream? Had we really been flying?

"Get up, lad. Today is the day you've been waiting for."

I stirred, reluctantly, then opened my eyes. As usual, the first thing I noticed was Zed's T-shirt: *Wo Ist Die Wasserstoff Farm?* I smiled dreamily, then said, "Just tell me Zed; did we really..."

"Fly? Indeed we did. There's nothing to it, once you get the hang of it. It's just a matter of forgetting it's impossible. Now hurry and get up, will you?"

What was the rush, anyway? The day was just beginning. "What is it, Zed?" I asked. "What's happening?"

Elusively, he said, "It's time, that's all. Your horse is waiting."

I pulled on my jeans and boots and threw my shirt over my shoulders. Outside, Mike was saddled and shifting his weight from one foot to the other, antsy to go. "Zed? What's going on here?"

There was a thinly veiled tone of anxiousness in his voice. "It's time for you to go back and tell the world what you've learned here, lad, but before you go there's one more secret you need to uncover. I'm sure you know what I mean."

"The interior. The strange machinations," I said, more to myself.

"You'd best be on your way. I'll see you in another time."

"What do you mean, Zed? I'll see you on my way out—won't I?"

"Sure you will," he said, though I knew it wasn't true. Bewildered and dismayed, I jumped on Mike and rode as fast as he could take me, toward the center of the Wasserstoff Farm, never looking back.

Through the orchards and the fields of grain, Mike ran like a horse on fire. He ran so fast that, until we reached the dwindling edge of the wheat, the wind was at our face. Now, no matter how much ground his feet chewed up—and his pace had hardly slackened—the wind was at our backs.

For another mile we ran, long after nothing grew from the ground but low grasses. The dust kicked up by Mike's hooves rushed ahead of us in tight swirls, and the noise grew deafening. Still Mike insisted on running hard and fast, long after I had lost heart in the matter. And I knew then what Zed had done to my horse that day in the mist, when we considered the pros and cons of the foreboding forces locked inside the atom.

Zed had charged Mike with a quest and my green-broke horse, who never seemed to pay me much mind, had paid heed to the wizard.

Despite the mounting heat and the worsening wind, Mike ran until we reached the heart of the Wasserstoff Farm. Sweat poured down my face and soaked my clothes. Mike was covered with a frothy white lather that was quickly becoming the color of dust.

When he could go no farther, Mike hit the brakes and whinnied in awe and confusion. Having reached his goal, it seemed he'd been released from Zed's spell and now found himself overcome by a host of threatening and unsavory sensory inputs.

As I studied what was before me from astride my increasingly agitated horse, the source of all the noise was now apparent: from a colossal fluted base, an enormous concrete chimney rose several thousand feet from the center of the Wasserstoff Farm. Two or three hundred feet up, a swarm of turbines screamed as hot air rushed past their gargantuan blades to exit through the chimney. So that was it: the Wasserstoff Farm was a giant solar-powered wind farm!

It all made sense now—why it was always hazy and breezy at the Wasserstoff Farm, and warm, even in late fall. I was in a greenhouse, covered by some sort of rigid, yet transparent, polymer; a polymer that may or may not have come from this world. Wherever the invisible roof had originated, it undoubtedly covered the largest greenhouse ever built. Cool air was drawn in through the deep crenellations in the top of the distant wall and heated by sunlight as it moved ever faster toward the center. The turbines were spun by a powerful convection current that moved air from the perimeter walls of the Wasserstoff Farm up through the massive chimney.

As I wiped the sweat from my face and spoke reassuringly to my horse, I pondered how many megawatts such a structure might produce. But before I could arrive at a number, there was an abrupt and alarming change in the surroundings. The wind—which had been steady at 35 or 40 miles per hour—grew to such intensity that it shook the turbines in their frames and rattled the roof high overhead, where it met the chimney near the level of the turbines. The very ground beneath our feet trembled, then began to roll in waves, as if it were liquid. I feared the place would soon come apart.

For Mike, the growing instability above, below, and all around us, meant it was time to boogie. In a Thoroughbred sort of way.

The colt spun around on his back feet and headed into the fierce wind as fast as his tired but powerful legs could carry us. It was all I could do to lower my head and shield my eyes as we were scraped and pelted by wind-borne sand and battered by flying foliage ripped free from the fruit trees far in front of us. I felt his muscles strain and push beneath the saddle as he fought his way out of the center of the maelstrom.

Though there was no way to gauge just how far or fast the colt had gone, it must have been a mile or better before the wind stopped and the ground ceased its undulations—suddenly and completely. It was as if someone had shut off the power to a Herculean wind tunnel and earthquake simulator. Nor was there any noise of whining turbines or air rushing through a 300-story chimney.

I raised my head to look about, then, and I couldn't believe what I saw. Or, rather, *didn't* see.

To my utter astonishment, the Wasserstoff Farm had vanished.

epilogue

★ ★ ★

Exhausted, we slowly made our way home over familiar ground. With the Wasserstoff Farm inexplicably gone, the terrain had reverted back to a typical mountain valley. The vast fields and orchards, all the structures, and every trace of the mighty perimeter wall, were gone. Head lowered, Mike trod along the single trail snaking through the center of it.

Had it ever been real? It was at once difficult to believe and impossible to deny.

It was late afternoon when my house became visible through the trees, and Mike instantly perked up his ears. He took off on a fast trot and headed straight for the barn. I unsaddled him, gave him some oats—which he greedily devoured—and made my way into the house.

How long had I been gone? Four days? Five, perhaps? So why did the clock on the table say it was Wednesday, November 10, the same day I'd left? Curious, I turned on my computer. A quick check of the satellite internet home page confirmed it: I'd been gone for only five hours.

Was I losing my mind?

Then why were the scratches and bruises I'd suffered on the day I discovered the Wasserstoff Farm now mostly healed? And why did I know so much more than I did before? You can't dream knowledge you don't have...can you?

The answer finally became clear—or as clear as it would ever be, at any rate—when I wandered into the kitchen for the pastrami sandwich and cold beer I'd been craving for...days?

The confirmation I needed was on the kitchen table—a T-shirt I was quite certain I didn't own. Written boldly across the front were the words, *Wo Ist Die Wasserstoff Farm?* It was the same shirt Zed had worn this morning. I didn't need a German dictionary to know what it meant. *Where is the Wasserstoff Farm?* Where, indeed.

Beneath the shirt was a note, written in the old wizard's practiced, flamboyant hand:

> Sorry for the hasty exit, lad, but it was time to pull up stakes and move on. To paraphrase a seaman's maxim, "the dimensional vortex waits for no one." Perhaps I'll be able to explain it to you, someday. If not, it'll give you something to ponder long after the seeds of the hydrogen revolution have sprouted and flowered.
>
> As for the final secret: the Wasserstoff Farm, you doubtless discovered, is a solar chimney, a city-sized convection generator, if you will. The turbine generators you saw were collectively rated at 200 megawatts and capable of fulltime operation. I wish I could say I invented it, but perhaps it's better that I didn't. If such things didn't already exist, you might have trouble convincing anyone it was possible.
>
> Zed
>
> PS: You will mention me in your book, won't you, lad?

Beside the note was a strangely luminescent ballpoint pen. Written on the barrel were the words: *Zedediah Pickett, Wizard Extraordinaire of The Wasserstoff Farm.*

Smiling broadly and staring into space, I played absently with the pen in my hand. Somehow, somewhere, I'd cross paths with ol' Zed again.

solving the mystery

A SOLAR CHIMNEY?

It's true. The original prototype was built in Manzanares, Spain, in 1981–82, and ran intermittently until 1989. Having proved the viability of the technology, the German Structural Engineering firm of Schlaich Bergermann und Partner is now building a much larger plant near the town of Mildura, Australia. The chimney will be nearly 3,160 feet *[one kilometer]* high, and the surrounding solar collection area will be 3 miles *[5 kilometers]* in diameter. The thirty-two turbines are expected to produce up to 500 gigawatt hours of electricity per year, enough to power 200,000 homes. For details, see the Vision Engineering website at: *www.visionengineer.com*. For a explanation of the concept, download "The Solar Chimney" pdf at: *www.math.purdue.edu/~lucier/The_Solar_Chimney.pdf*. Or better yet, read the book, *The Solar Chimney*, by Jorg Schlaich, himself.

This prototype solar chimney plant in Manzanares, Spain operated for over 15,000 hours from 1982–89. It ran an average of 8.9 hours per day from 1986–89, with only one person needed for supervision. *Photo courtesy of Schlaich Bergermann und Partner.*

ACKNOWLEDGMENTS

At last. The months of grueling research and sleepless nights are over. No more 4:00 a.m. flashes of inspiration and subsequent calls to the keyboard; I can sleep past sunrise without self-recrimination. The book is, as they say, "in the can." But now I'm faced with the *really* hard part—trying to figure out how to properly thank all the people who worked on the sidelines to make *Hot Stuff Cool Science* possible.

First, I have to thank my wife and soul mate, LaVonne. About halfway through the first draft, I began to think this might be the one book where my hours of labor would exceed hers. I was wrong. Her dedication to the project was unwavering, and her editing, design and layout skills were, and are, unequalled. She bravely questioned my judgment in all the places it needed questioning, even if I was in no mood to have my judgment questioned. She was always there to encourage me when the going got rough, and she pulled me back to the ground whenever I got too cocky. And she did it all with a sense of diplomacy beyond my muted understanding of such matters. Thanks, Honey—you're the one.

Dr. Ronal Larson, Chair-elect of ASES, read every draft of the manuscript, and routinely deluged me with ideas, corrections, and leads on new references. He was relentless. No one else I know—myself included—would have taken the manuscript on a vacation to the Galapagos Islands, but Ron did. Needless to say, many wonderful things in this book would not be there, if not for Ron's insistence. Thanks, Ron, for all your invaluable help in making this book a reality.

My books would not be nearly as fun without Sara Tuttle's amusing drawings. Certainly, no one other than me would know what Zed looks like, if not for Sara's whimsical portrayals. So well does she apply her talents, in fact, that I can easily overlook the liberties she took when drawing my moustache. Many thanks, Sara—we'll have to do this again, soon.

Dr. Doug Larson, owner of FuelCellStore.com (and no relation to Ron), waved a copy of Kenn Amdahl's outrageously fun book, *There Are No Electrons,* in front of my face at NREL one day and told me the world needed a book that did for hydrogen what Kenn's book did for electronics.

Without pausing to ponder what I was getting myself into, I said I'd give it a shot. It's all your fault, Doug.

For reading all or part of the manuscript and making a boatload of good suggestions and valuable comments, I thank Linda Masterson, friend and fellow writer; Steve Clark of the Institute of Ecolonomics; Danny Day, owner of Eprida; Raymond Atwood, award-winning high school science teacher; and Steve Iwanicki, Jr., for giving me the much-needed perspective of an intelligent—yet healthily-distracted—high school senior.

Dennis Weaver is the founder of the Institute of Ecolonomics and an actor I have admired all my life. He's also a man who knows good horseflesh when he sees it. Somehow, he found time in his hectic schedule to read the manuscript and give it a hearty thumbs-up. Thanks, Dennis!

The hydrogen revolution would be science fiction without help from the countless corporations, institutions, organizations and individuals who supplied concepts, research papers, and photos for this book. I thank you, one and all.

The internet abounds with search engines, and through the course of writing *Hot Stuff Cool Science* I must have tried them all. Among them, Google stands alone. Thanks guys—you made my job a whole lot easier.

And finally, I have to offer my undying gratitude to my dear friend and mentor, Zedediah Pickett. Some will say you're not real, Zed, but we know better, don't we?

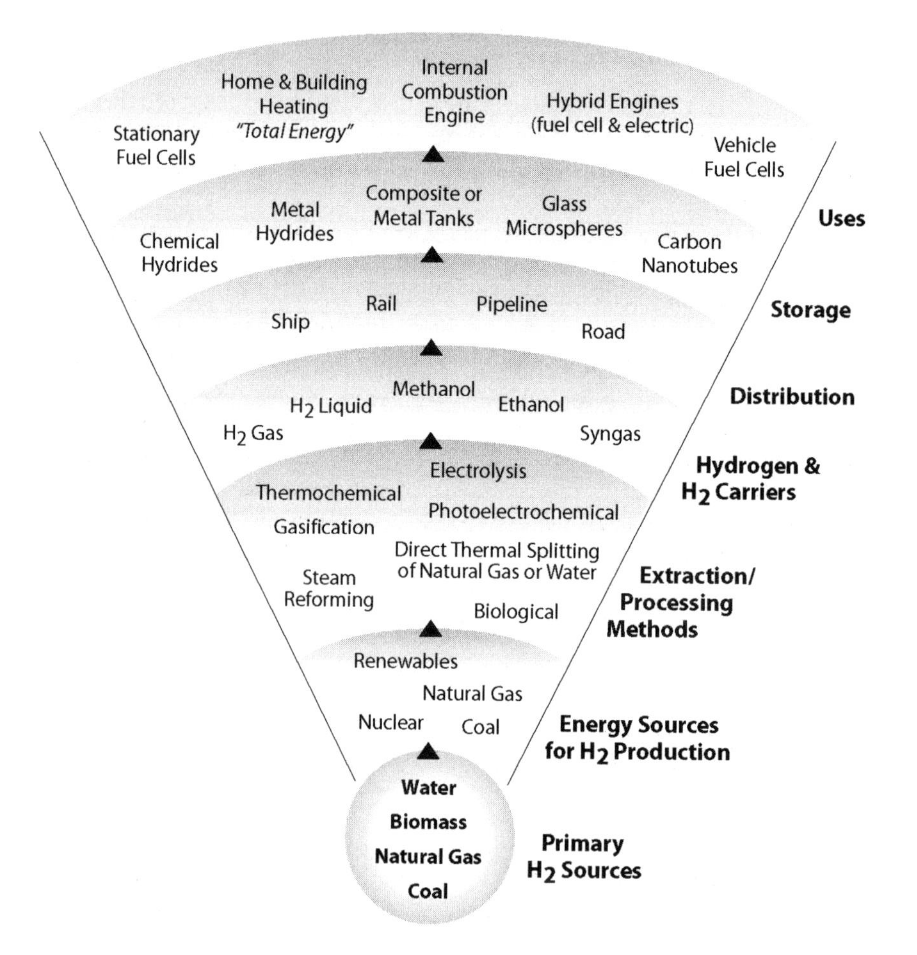

Moving Towards a Hydrogen Economy

INTERVIEW WITH RONAL LARSON, PhD

Vice-Chair of the American Solar Energy Society

AUGUST 2004

Dr. Ron Larson is a retired former Professor (E.E. at Georgia Tech) and former Principal Scientist at the Solar Energy Research Institute (now the National Renewable Energy Laboratory - NREL). His solar energy activities began in 1973 as the first IEEE Congressional Fellow, working on the first solar legislation passing the House Science Committee. As an American Solar Energy Society (ASES) board member, he has been responsible for Membership and Chapters activities, and is now Vice-Chair (Chair-Elect) of ASES. In 1996, Dr. Larson was instrumental in establishing the Colorado chapter of ASES, of which he remained active on regulatory and legislative issues. He has also served as the volunteer coordinator of an Internet list called "stoves" (for developing countries, under the sponsorship of the Renewable Energy Policy Project). His *stoves* and international interests grew out of leading a USAID project in Sudan in the early 1980s. Dr. Larson is active in the growing hydrogen community, and his strongly-held views on the role of renewables in the coming hydrogen economy are widely known and respected. Information about Dr. Larson's and ASES' involvement in the promotion of a hydrogen economy can be found at *http://www.ases.org/hydrogen_forum03/h2home.htm*.

REX: You've read the book, Ron. Despite the many hurdles that remain, Zed is clearly gung-ho on the idea of the hydrogen economy. Do you share his enthusiasm?

RON: I have enthusiasm for almost everything Zed has said, but I'm afraid that Zed may not be looking enough at the hurdles—which are mostly economic, but also involve efficiencies.

Rather than hydrogen, for instance, we may initially see a heavy use of batteries because of their superior efficiencies and economics. Or we may see biofuels—the ethanol and methanol options Zed discussed—but used perhaps with a hybrid car with a much larger battery bank which can be recharged from the grid, or even with a home PV and wind system.

Because inefficiencies and economics are the two big drawbacks for the hydrogen economy, in a future book I hope Zed will show us how to perform these miracles as well as the physical ones he does so well in this book. But however we go about it—batteries, biofuels, or hydrogen—the future is surely with renewables and it was great to share Zed's enthusiasm and optimism there.

REX: *Let's talk a bit about ethanol and methanol—the "alcohol sisters" as Zed fondly refers to them. When made from biomass, they show great promise as fuels capable of actually decreasing atmospheric CO_2, since it's relatively easy to capture most of the CO_2 given off during processing. Add to that the fact that ethanol and methanol are excellent room temperature hydrogen carriers and they appear to be ideal fuels. How big of a role do you think they deserve in the coming hydrogen economy?*

RON: I like all parts of ethanol and methanol, except the possible consequent need for rapidly declining surplus land for an already too-large global population. In such a world, we will certainly have to move to wind and direct solar conversion systems that can convert incoming solar radiation more efficiently. But this need not preclude both ethanol and methanol as energy carriers, if we can emphasize the use of agricultural waste materials.

And yes, you can capture the CO_2 during processing, but then something has to be done with it. This means either sequestration underground or the charcoal soil-enrichment process—based on Danny Day's work—that Zed discussed in chapter 10. Otherwise, alcohol fuels only slow the rate of increase of atmospheric CO_2.

REX: *Hybrid cars have shown a lot of promise. Might they be used to make better use of a finite alcohol production capacity?*

RON: A different type of hybrid car—one with batteries capable of a 50–60 mile range that can be plugged into an electrical source—may cut current transportation fuel consumption by 90 percent. The U.S. could sustain such a liquid fuel demand with biofuels and still feed much more than our own population. Alcohol fuels used this way could be a tough competitor for the hydrogen option, not only because they can be integrated well into our existing electric and liquid fuel distribution systems, but also because they offer rural America and other countries a chance for much needed economic development. This type of hybrid option could use either internal-combustion engines, or hydrogen or alcohol fuel-cells, depending on how the latter progresses.

REX: *What about timing? Depending on who you talk to, the beginnings of the hydrogen economy are as close as a few years, or as far away as several decades. And yet it appears all the pieces to the puzzle are tantalizingly close to fitting together. How far away from a hydrogen economy are we, in your opinion?*

RON: As someone who once thought it quite possible to achieve the Carter Presidency goal of 20 percent renewables by 2000, I am now dubious of anything faster than a one percent change per year. So I have to say it may be decades away. Recent technological revolutions with rapid growth rates such as the computer or cell-phone revolutions are great models, but they were filling voids. I cannot be optimistic with the low rate of introduction that is now in process.

REX: *What do you see as the main stumbling block here? Certainly more research is needed—indeed, it will always be needed—but how much more R&D is required before we can proceed to get our feet wet?*

RON: The main stumbling block, I believe, is the inertia of changing any major economic system in the absence of a major national crisis. Of course we need more R&D, but we DO have to get started with H_2 introduction in an ever-increasing commercial scale. I suppose we can't avoid obtaining much of the H_2 from natural gas initially, but simultaneously we must make renewables grow at an even faster rate, even if we don't use them to produce H_2 in the beginning. The global climate change and resource depletion problems demand nothing less.

REX: *In the final chapter Zed alluded to a need for strong leadership to get us started on the right path. Is he right?*

RON: Yes—we do need strong leadership. Thankfully, that leadership has already appeared in Europe and Japan. Until we have a real crisis, however, I fear that the entrenched conventional fuels—coal, oil, gas, and nuclear—will find ways to keep both the solar and hydrogen transitions going slowly in the United States.

REX: Coal is interesting. It appears to be the proverbial double-edged sword. Despite its bad reputation, coal—more than any other conventional fuel—may be a cheap and easy way to get the hydrogen ball rolling. What are your thoughts about coal?

RON: Until recently I felt there was no way that coal could contribute significantly toward fulfilling our energy demands for the next one hundred years, due to its huge contribution to global warming and climate change. Now, after conversations with Dr. Bob William of Princeton University, who has looked at the sequestration issue deeply, I am less certain on what can and should be done. He notes correctly that coal is cheap and relied upon in China and India. These developing countries are justifiably interested in advancing their economies as we have, but they are not going to listen to the United States who is ignoring climate change issues. So I now am switching to the position that we must move quickly to studies of sequestration practicality, while also aggressively developing (my preferred) renewables. Dr. William feels that, with gasification, sequestration costs could add a relatively small percentage price increase—perhaps 1.5 cents per kWh. But there is not yet any proof that sequestration several kilometers down—in suitable strata—can be done with assured retention, and too few rigorous tests are being planned. As a nation, the U.S. should move rapidly and then give the technology away to coal-using, developing countries. But we in the hydrogen community have little power to force this. Hopefully your book can help promote the need for much quicker certainty about sequestration.

REX: Considering your position on renewable energy, I'm sure you have misgivings about all this interest in coal and CO_2 sequestration. True?

RON: True. It seems to me that the global transition to renewables is slowing down, largely because of the abundance of coal and the belief that long-term sequestration is possible. But your question also goes to efficiencies of production and the issue of hydrogen versus electricity as an energy carrier. As you have discussed with biomass, one can convert from coal to hydrogen without the intermediate electricity step, so there will be enormous pressure to go in that direction as well—further slowing down the inevitable transition to renewables. The solution to this dilemma is a larger Federal R&D budget, as this sequestration issue needs earnest and immediate attention.

REX: But if the prospects for hydrogen from clean coal should pan out?

RON: This is a difficult question. I cannot now say with certainty that the clean, economic production of hydrogen from coal, or the sequestration of CO_2, is impossible. This uncertainty will slow us down considerably until it is clear that enough hydrogen—or electricity—can be produced more economically from renewables than from coal, a possibility that is already true for wind, at least in some locations. And the problem is further compounded by the lack of proof that we face serious near-term problems with both oil and natural gas supplies. Too many people are optimistic that we have plenty of both.

REX: What about nuclear? While I don't see any real passion for it, there seem to be many who will grudgingly accept it, if it hastens the transition to a hydrogen economy. In your opinion, would nuclear get us there any faster, and if so, would it be worth it?

RON: I'm not so sure that "many...accept it." My perception is that nuclear energy's main problem has been this very lack of acceptance—a NIMBY *[not in my backyard]* response, based on safety concerns. Hydrogen production from nuclear sources is certainly and justifiably being promoted in large part because of its global climate change advantages, but I doubt that is enough to speed a nuclear-hydrogen future. My main concerns are the high and uncertain costs of nuclear energy plants, potential core meltdowns (due to design, material, operational and terrorism-related flaws), long-term fuel sufficiency, waste disposal concerns, and weapons proliferation, this latter making worldwide acceptance almost impossible. This is just too long of a list to think a nuclear-hydrogen future would be either fast or at all "worth it." Fortunately, for those of us promoting renewable energy, only the cost issue arises.

REX: Okay. Let's end this where we started. At the beginning of the interview you said that inefficiencies and economics had quelled your enthusiasm, yet in many instances these seem to be facts of life that we will one day be forced to live with. Your final thoughts?

RON: Well, we can certainly tolerate inefficiencies better with renewable energy inputs than with fossil fuels. If solar or wind energy is captured and then inefficiently used or converted there is no great harm, as the input energy could not be saved anyway, and it causes little negative impact. With coal, on the other hand, an inefficient conversion adds quite negatively to climate change as well as the many other negative impacts arising from coal use today. We are in agreement on this I suspect, as I believe Zed made this same point. But inefficiencies also lead to higher costs, and cost minimization will drive our economy for years. My concern is much more on the practical economics of a conversion to hydrogen—and especially of the problems associated with establishing a totally new delivery system.

I'm hoping that we in the United States can tackle these issues as well as leaders in Europe and other parts of the world already have.

REX: Thanks, Ron...

RON: Uh, Rex, could I ask you a question, now?

REX: Of course, Ron.

RON: Do you suppose I could meet Zed, someday? I'm sure we could find a lot to talk about.

REX: No doubt you could. I'll see what I can arrange.

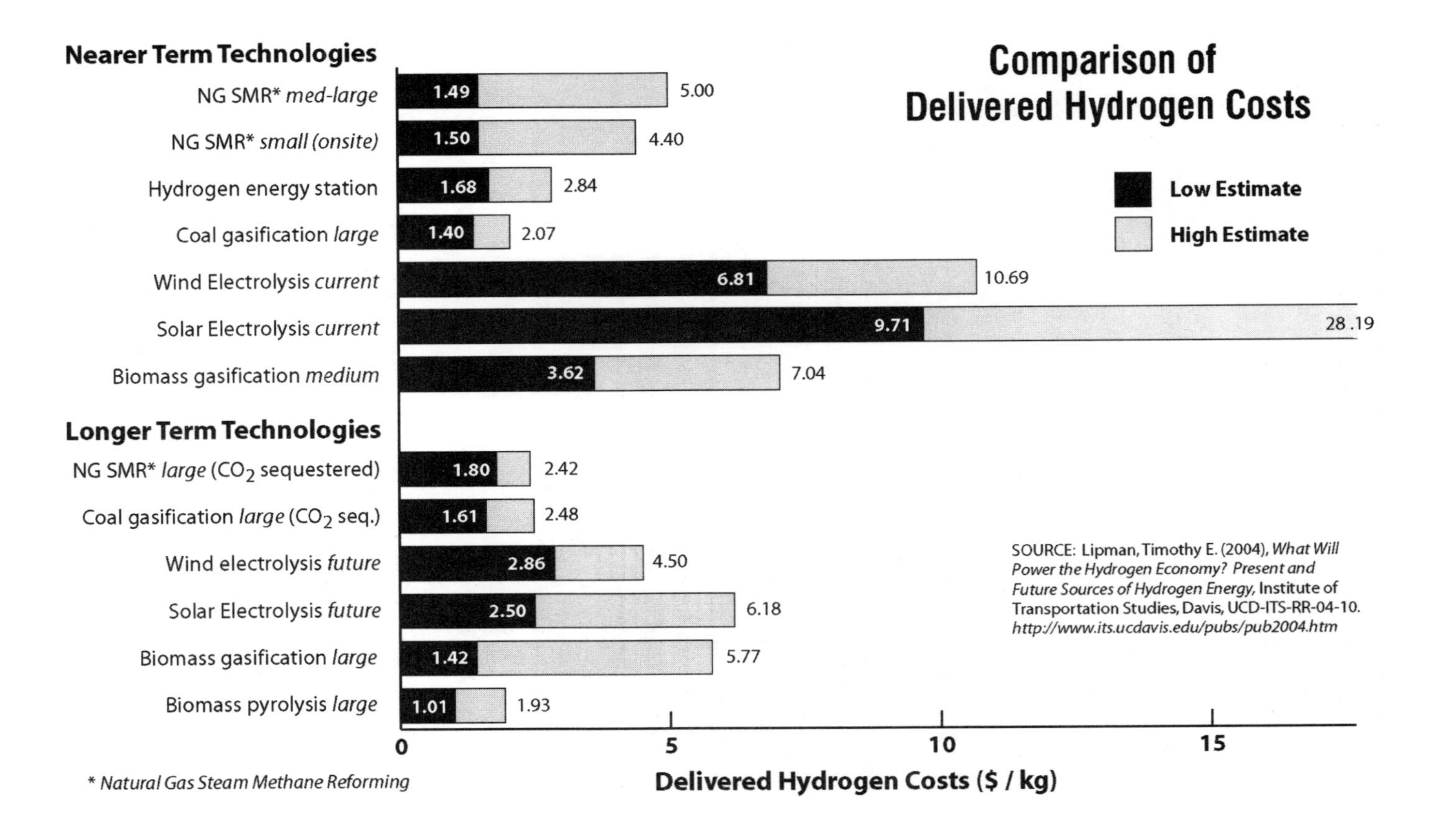
Comparison of Delivered Hydrogen Costs
Low Estimate
High Estimate
Nearer Term Technologies
NG SMR* med-large
1.49
5.00
NG SMR* small (onsite)
1.50
4.40
Hydrogen energy station
1.68
2.84
Coal gasification large
1.40
2.07
Wind Electrolysis current
6.81
10.69
Solar Electrolysis current
9.71
28 .19
Biomass gasification medium
3.62
7.04
Longer Term Technologies
NG SMR* large (CO_2 sequestered)
1.80
2.42
Coal gasification large (CO_2 seq.)
1.61
2.48
Wind electrolysis future
2.86
4.50
Solar Electrolysis future
2.50
6.18
Biomass gasification large
1.42
5.77
Biomass pyrolysis large
1.01
1.93
0
5
10
15
Delivered Hydrogen Costs ($ / kg)
SOURCE: Lipman, Timothy E. (2004), What Will Power the Hydrogen Economy? Present and Future Sources of Hydrogen Energy, Institute of Transportation Studies, Davis, UCD-ITS-RR-04-10. http://www.its.ucdavis.edu/pubs/pub2004.htm
* Natural Gas Steam Methane Reforming

CONVERSION FACTORS

Since most of the world does not think in English units of measure like Zed and I do, I've made many conversions to metric throughout the book. There are, however, many places—particularly within dialog—where parenthetical conversions would have been unforgivably distracting. So, for those of you wishing to "translate" those values from English to metric, here is a list of the conversion factors used.

HEAT

Btu x 1.054 = kilojoules (kj)	*kj x 0.9485 = Btu*
Btu/pound (lb.) x 2.326= kj/kg	*kj/kg x 0.4299 = Btu/lb.*
Quadrillion Btu (quads) x 1.054= exajoules	*exajoule x 0.9485 = quads*

PRESSURE

While the average American thinks strictly in pounds per square inch (**psi**) everyone else on the planet tends to think in **pascals** (1.0 newtons per square meter), **kilopascals** (1,000 newtons per square meter), **megapascals** (1,000,000 newtons per square meter), or **bars** (10 megapascals). Atmospheres are also sometimes used and, depending on if you are talking about a technical atmosphere or a standard atmosphere, it will be just a pinch more or less than a bar. In any event, everything converts easily to everything else, except for the psi which is the oddball of the bunch. But, unless you demand exactness, you can mentally figure 1000 psi = 7 megapascals, or 70 bars. The exact values are:

1000 psi = 6.895 megapascals	*1 megapascal = 145 psi*
1000 psi = 68.9476 bars	*1 bar = 14.5 psi*

AREA

Square feet x 0.0929 = square meters	*Sq. Meters x 10.76 = Sq. Feet*
Acres x 0.4047 = hectares	*Hectares x 2.471 = Acres*
Square miles x 2.59 = square kilometers	*Sq. Km x 0.3861 = Sq. Miles*

LINEAR (DISTANCE)

Inches x 2.54 = centimeters	*Centimeters x 0.3937 = inches*
Feet x 0.3048 = meters	*Meters x 3.28 = feet*
Yards x 0.9144 = meters	*Meters x 1.093 = yards*
Miles x 1.609 = kilometers	*Kilometers x 0.6213 = miles*

SPEED

Miles per hour (mph) x 0.447 = meter per sec (m/sec)

Meters per sec (m/sec) x 2.2369 = mph

Miles per hour (mph) x 1.609 = kilometers per hour (kph)

Kilometers per hour (kph) x 0.6213 = mph

WEIGHT

Ounces (oz.) x 28.3495 = grams (gm)	*Grams x 0.035 = ounces*
Pounds (lb.) x .4536 = kilograms (kg)	*Kilograms x 2.2 = pounds*
Tons x 0.907 = tonnes (metric tons)	*Tonnes x 1.1023 = tons*

LIQUID AND VOLUME

Gallons (gal) x 3.7854 = liters (l)	*Liters x .2641 = gallons*
Cubic feet (ft^3) x 0.02832 = cubic meters (m^3)	*m^3 x 35.3147* = ft^3

TEMPERATURE

Degrees Celsius = (degrees Fahrenheit – 32) x .55556 approx.

Degrees Fahrenheit = (degrees Celsius x 1.8) + 32

THE REALLY BIG NUMBERS

Million	6 - 0's	10^{6}	1,000,000
Billion	9 - 0's	10^{9}	1,000,000,000
Trillion	12 - 0's	10^{12}	1,000,000,000,000
Quadrillion	15 - 0's	10^{15}	1,000,000,000,000,000
Quintillion	18 - 0's	10^{18}	1,000,000,000,000,000,0000
Sextillion	21 - 0's	10^{21}	*you get the idea...*
Septillion	24 - 0's	10^{24}	
Octillion	27 - 0's	10^{27}	
Nonillion	30 - 0's	10^{30}	
Googol	100 - 0's	10^{100}	

RESOURCE WEBSITES

Additional references listed in Technistoffs in each chapter.

American Methanol Institute *www.methanol.org*
American Solar Energy Society *www.ases.org*
American Wind Energy Association *www.awea.org*
Ballard *www.ballard.com*
BMW *www.bmw.com/generic/com/en/fascination*
Ecolonomics *www.ecolonomics.org*
Energy Efficiency and Renewable Energy *www.eere.energy.gov*
European Wind Energy Association *www.ewea.org*
Ford Motor Company *www.ford.com/en/innovation*
4Hydrogen *www.4hydrogen.com*
FuelCellStore.com *www.fuelcellstore.com*
Fuel Cell Markets *www.fuelcellmarkets.com*
Fuel Cells 2000 *www.fuelcells.org*
General Motors *www.gm.com/company/gmability* (select Technology)
H2CarsBiz *www.h2cars.biz*
How Stuff Works *www.howstuffworks.com*
HyWeb *www.hydrogen.org*
International Association for Hydrogen Energy *www.iahe.org*
International Solar Energy Society *www.ises.org*
National Fuel Cell Research Center *www.nfcrc.uci.edu*
National Hydrogen Association *www.hydrogenus.com*
National Renewable Energy Laboratory *www.nrel.gov*
Plug Power, Inc. *www.plugpower.com*
Quantum Fuel Systems Technologies Worldwide, Inc. *www.qtww.com*
Renewable Fuels Association (ethanol industry) *www.ethanolrfa.org*
Shell Hydrogen *www.shell.com/home/Framework?siteId=hydrogen-en*
U.S. Fuel Cell Council *www.usfcc.com*
World Fuel Cell Council *www.fuelcellworld.org*

GENERAL BIBLIOGRAPHY OF BOOKS

Additional references listed in Technistoffs in each chapter.

Asimov, Isaac, *Building Blocks of the Universe*, Revised Edition. New York: Lancer Books, 1966

Atkins, P. W., *Molecules*. New York: Scientific American Library-HPHLP, 1987

Gipe, Paul, *Wind Power for Home and Business*. Post Mills, VT: Chelsea Green Publishing, 1993

Glover, Thomas J., *Pocket Ref*. Littleton, CO: Sequoia Publishing, 2001

Gundersen, P. Erik, *The Handy Physics Answer Book*. Farmington Hills, MI: Visible Ink Press, 1999

Hawking, Stephen, *The Universe in a Nutshell*. New York: Bantam Books, 2001

Heiserman, David L., *Exploring Chemical Elements and their Compounds*. Blue Ridge Summit, PA: TAB Books, 1992

Hoogers, Gregory, et al., *Fuel Cell Technology Handbook*. Boca Raton, FL: CRC Press, 2003

Peavey, Michael A., *Fuel From Water*. Eleventh Edition. Louisville, KY: Merit, Inc., 2003

Rhodes, Richard, *The Making of the Atomic Bomb*. New York: Simon and Schuster, 1986

Schlaich, Jorg, *The Solar Chimney*. Baden-Württemburg, Germany: Edition Axel Menges, 1996

Strauss, Stephen, *The Sizesaurus*. New York: Kodansha America-Key Porter Books, 1995

SUGGESTED READING

Natural Capitalism, Paul Hawken, Amory B. Lovins, Hunter L. Lovins

Power to the People, Vijay V. Vaitheeswaran

There Are No Electrons, Kenn Amdahl

Tomorrow's Energy, Peter Hoffman

INDEX

Italics numbers represent illustrations, graphs or photos

LIST OF ILLUSTRATIONS

ABOUT THE AUTHOR

Rex A. Ewing is a writer who loves a challenge, especially when it comes wrapped up in numbers and science that beg to be explained. When people in the hydrogen industry and the book world asked him to write a book about hydrogen and fuel cells for the non-scientist, he couldn't resist the opportunity to take dry—some would say incomprehensible—science and craft a page-turner that reads like a best seller.

From his solar- and wind-powered studio in the Colorado Rockies, Ewing has also written *Power With Nature: Solar and Wind Energy Demystified*, a best-selling book for homeowners wanting to learn about renewable energy; and *Logs, Wind and Sun: Handcraft Your Own Log Home...Then Power It With Nature*, a handbook for aspiring log home owners who want to live off-grid. He is also a regular contributor to *Log Homes Illustrated* magazine.

Before moving to the mountains to concentrate on his writing, Ewing spent several years as CEO of a well-respected equine nutrition firm, where he formulated and marketed numerous nutrition products worldwide. In 1997, he wrote a best-selling book on horse nutrition: *Beyond the Hay Days: Refreshingly Simple Horse Nutrition*. His articles frequently appeared in popular equine publications. The updated and expanded 2nd edition of *Beyond the Hay Days* was released 2003.